"This book is written with the intent to encourage development of policies to prevent and limit the spread of future outbreaks by not only educating policymakers, but engaging a broader audience. Harvell makes a plea for action, citing the need for surveillance of farmed species like salmon, shrimp, oysters, and abalone, where disease outbreaks also pose threats for wild populations."
—*Fisheries*

"Part memoir and part science exposé, *Ocean Outbreak* is a recipe for launching into the unknown, providing a road map showing how one person can marshal the world of science to win against a global peril."
—Steve Palumbi, author of *The Evolution Explosion: How Humans Cause Rapid Evolutionary Change*

"A marine medical whodunit, where the patients include sea stars facing odds worse than humans did during the Black Death."
—Nancy Knowlton, author of *Citizens of the Sea*

"*Ocean Outbreak* brilliantly lays out the risk of disease and smart plans for improving ocean health."
— Ted Danson actor, activist, and founding board member of Oceana

Ocean Outbreak

The publisher and the University of California Press Foundation gratefully acknowledge the generous support of the Ralph and Shirley Shapiro Endowment Fund in Environmental Studies.

Ocean Outbreak

*Confronting the Rising Tide of
Marine Disease*

DREW HARVELL

UNIVERSITY OF CALIFORNIA PRESS

University of California Press
Oakland, California

© 2019 by C. Drew Harvell

First Paperback Printing 2021

Library of Congress Cataloging-in-Publication Data

Names: Harvell, C. Drew, 1954– author.
Title: Ocean outbreak : confronting the rising tide of
    marine disease / Drew Harvell.
Description: Oakland, California : University of
    California Press, [2019] | Includes bibliographical
    references and index. |
Identifiers: LCCN 2018049384 (print) | LCCN 2018051515
    (ebook) | ISBN 9780520296978 (cloth : alk. paper) |
    ISBN 9780520382985 (pbk.: alk. paper) |
    ISBN 9780520969506 (ebook)
Subjects: LCSH: Marine ecosystem health. | Corals—
    Diseases. | Abalones—Diseases. | Salmon—
    Diseases. | Starfishes—Diseases.
Classification: LCC QH541.5.S3 (ebook) | LCC QH541.5.S3 H37
    2019 (print) | DDC 577.7—dc23
LC record available at https://lccn.loc.gov/2018049384

Manufactured in the United States of America

27 26 25 24 23 22 21
10 9 8 7 6 5 4 3 2 1

To cherish what remains of the Earth and to foster its renewal is our only legitimate hope of survival.

Wendell Berry

Contents

Illustrations

Preface

In the aftermath of a devastating outbreak that brought ocean disease to the public eye in 2015, I was invited to organize an Ocean Outbreaks Day during which members of Congress would be briefed on growing threats to organisms in our coastal waters. Our team of four scientists traveled to Washington, DC, and gave talks on diseases that were plaguing reefs in Florida and Hawaii, impacting oyster fisheries in North Carolina and elsewhere along the eastern seaboard, threatening lobster fisheries in Long Island and Maine, and causing harm to salmon fisheries in Washington and abalone in California. Throughout our presentations we emphasized how disease outbreaks in US coastal waters had economic consequences.

There was keen bipartisan interest in ocean health that day. However, the questions we received from lawmakers and their staffers revealed to us a large information gap. Those from ocean states were knowledgeable about some issues affecting the sustainability of our oceans. They were aware of overfishing and the need for a restructuring of fisheries management. They knew about nutrient pollution and eutrophication and the dangers of polluting the oceans with both large plastics and insidious micro-plastics. They had concerns about the emerging problem of ocean acidification and its effects on organisms that make calcareous shells. Even if the impacts still seemed far off, staffers and legislators from Hawaii and Florida were aware that warming temperatures were bringing devastation to coral reefs around the world. But few people in the hearing rooms were more than vaguely aware of the rolling tide of new disease outbreaks that is affecting marine organisms around the world. They did not know that warmer and more polluted ocean waters are allowing many infectious pathogens to thrive while at the same time weakening marine creatures' abilities to withstand disease.

.   .   .

Most policymakers we briefed had never thought about what would happen to their state's economies if warming oceans made for sicker marine organisms. They were clearly concerned, and it was motivating to see them respond constructively. Their excitement was palpable as staffers brainstormed about new legislation focused on ocean health. Many of them indicated their support for a new bill introduced by Congressman Dennis Heck of Washington, the Emergency Marine Disease Act (which ultimately failed to pass).

In that moment, I realized the urgent need for scientists to communicate to citizens and policymakers what we are learning about marine disease. We need to tell the stories of what we discover in the lab and the field. We need to translate our data and findings into forms that make the public more aware of the seriousness of the threats that ocean pathogens pose to our food supplies, economies, livelihoods, and health. We need to propose ways to protect the health of the oceans and better manage our impacts. That was the moment I resolved to do my part in this effort. After much reflection, I decided that a book detailing my experiences with four recent and serious disease outbreaks might be the best way to raise public awareness of the threat of marine diseases and motivate others to take action.

It takes a village to write a book like this and I am grateful to my friends, colleagues, and family members for their help and encouragement. My son Nathan, himself a writer, helped with hours of editing and ideas in the formulation, proposal writing, and early versions of the book. My daughter Morgan commented on the corals chapter. My husband Chuck listened patiently, advised on oceanographic issues, and helped brainstorm titles and frame ideas.

Substantive early help came from Steve Palumbi and Peter Sale, who read and advised on an early version. Developmental editor Eric Engles read every word twice and was a constant force in helping me translate my ideas from the ivory tower to a place closer to real lives.

The people in my research lab at Cornell have been mainstays of help, guidance, and fact-checking. Phoebe Dawkins edited, suggested good ideas and new wording, fact-checked, prepared literature lists, corrected spelling, prepared figures, and advised on maps, all with genuine enthusiasm and unceasing encouragement. Putu Dawkins drew the maps. Allison Tracy read, revised, and made suggestions on several chapters, including those about corals and salmon and the final chapter. Morgan Eisenlord helped with the starfish chapter and Olivia Graham with the final chapter. Joleah Lamb provided help on both the corals chapter and the final chapter. Cornell undergraduates in my research apprenticeship semester also commented on some sections.

My colleagues were generous with their time, helping me with fact-checking and related matters. Colleen Burge reviewed the chapters on abalone and starfish. Ian Hewson reviewed the starfish chapter; Carolyn Friedman and Brian Tissot the chapter on abalone; and Maya Groner, Paul Hershberger, and Kim Sundberg the chapter about salmon. At the eleventh hour, Scott Schwinge and Joe Gaydos read the entire book and helped weed out many extra references.

I am grateful to the team at the University of California Press: Kate Marshall, for supporting the concept and the book from the beginning; Dore Brown, for shepherding the book through the production process; and Jan Spauschus, for thoughtful copy editing.

I appreciate National Science Foundation funding of our Ocean Health Research Coordination Network. *Ocean Outbreak* communicates themes developed over the five years of that project and the views of the many scientists involved. Funding from the Atkinson Center at Cornell, the Nature Conservancy, and the Environmental Defense Fund facilitated the work in Indonesia and Myanmar.

CHAPTER ONE

What Rises with the Tide?

It is the microbes that will have the last word.

Louis Pasteur

You always remember the moment something bad turns big. For me, a bad situation assumed epic proportions in December 2013 when I was at the Nature Conservancy's All Science Meeting in California. I received an email from my colleague Pete Raimondi saying that thousands of starfish from at least ten species were dying fast in the waters around Monterey, California. I already knew that a species of giant sunflower starfish (*Pycnopodia helianthoides*) was dying catastrophically hundreds of miles to the north, not far from my home on San Juan Island in northwest Washington State. I had seen underwater photographs taken by Neil McDaniel near Vancouver, Canada, showing a disaster unfolding in the deep canyons there. In photos taken on October 19, the rock cliffs were covered with healthy-looking stars. In photos taken only ten days later, all that was left were hundreds of dead bodies piled on the sea floor beneath the cliffs (see figure 1). We had assumed this was a localized event affecting a single species, like others we had seen. Doubts about the geographic restriction of the starfish die-off had surfaced,

Figure 1. Mass mortality of the sunflower star over two weeks in British Columbia, 2013. Photos by Neil McDaniel.

however, when we learned in November, only a few weeks before the Nature Conservancy's meeting, that the Vancouver, Seattle, and Monterey aquariums had each lost hundreds of stars in their tanks. Against this background, Pete's email signaled something much more worrisome than a local die-off of one kind of starfish. It seemed we were seeing the beginnings of a disease outbreak that could end up affecting starfish along the entire Pacific coast.

In addition to being concerned about the broad geographic extent of the outbreak, I was worried that it involved starfish, and not just one species but many. Starfish may seem innocuous and almost inanimate given the glacial pace at which most species move, but they are the lions of our seascape. Even a few starfish can control the structure and composition of the surrounding ecosystem by eating huge numbers of the mussels and oysters that would otherwise dominate. Observing ochre stars preying on mussels and the changes that occurred when he removed them, experimentally, from his study areas on Tatoosh Island caused Bob Paine, one of my mentors, to invent the term *keystone species*. Added to that, the West Coast of the United States has a bright medley of different starfish species, second only to Australia for temperate waters. There are approximately eighty species described in the eminent marine biologist Eugene Kozloff's key for Puget Sound and the San Juan Islands in Washington. The catastrophic loss of not only ochre stars but other species as well over a broad swath of ocean could have a domino effect on ocean ecology, causing a cascade of changes that might ultimately impact animals—such as abalone and salmon—that humans depend on for food.

I left my session early and called Pete to hear firsthand what he and his fellow divers had seen in Monterey. He told me that they had watched the giant sunflower stars die first: they lost strength in their tube feet, their arms tore off and crawled away from their bodies, and their organs spilled out, leaving the stars to fall off the rock walls and docks. A week later, they had watched the same thing happen to other species. Sun stars (*Solas-*

ter sp.), rainbow stars (*Orthasterias koehleri*), giant pink stars (*Pisaster brevispinus*), giant stars (*Pisaster giganteus*), mottled stars (*Evasterias troschelii*), vermillion stars (*Mediaster aequalis*), and bat stars (*Patiria miniata*) all began to die rapidly in high numbers. One of the last to go was the ochre star (*Pisaster ochraceus*). The sea bottom, Pete told me, was littered with dead, decaying starfish arms and bodies, with crabs picking at them. The only species left gripping the rocks were the leather star (*Dermasterias imbricata*) and the blood star (*Henricia* spp).

It took me a moment to get past the gruesomeness of Pete's descriptions and assess their import. Many different species were dying over a very wide geographic range, from Vancouver all the way to Monterey on California's central coast. Almost all the most common starfish species were affected. Captive aquarium populations were being hit as badly as wild ones. They were dying rapidly, and few if any were surviving. It seemed unlikely that the culprit was some kind of horrific new coast-wide pollution problem—the die-off was too widespread. I had to conclude that we were facing a new marine epidemic, a disease that was killing an astonishing number of different species at a blistering pace and with a vast geographical reach.

The Nature Conservancy meeting turned hectic for me after I talked with Pete. I tried to attend sessions but my phone was ringing almost non-stop. One call was from Katie Campbell, a broadcast reporter with KCTV in Seattle. She wanted help with a television news story about all the dead starfish washing up on the beaches around the city. She had been told to call me because I lead a government-funded research network that specializes in ocean disease outbreaks and I teach an ecology of infectious disease course at the University of Washington marine lab near

Seattle. I filled her in on what I knew about the event. When we finished our conversation, I thought, "What am I doing at this meeting in California when stars are dying in large numbers in my home waters?" The die-off was unfolding in full public view all over our beaches and people were upset and concerned. I checked my tide table and saw that if I made it to Seattle the next day I could catch a low tide at around 8:00 pm. If I went to the right place, I could immediately see how the outbreak looked and begin collecting data.

Soon I was back on the phone, talking with Laura James, who called to tell me stars were dying at her favorite dive site, Cove 1 at Alki Beach, right near downtown Seattle. Cove 1 is surprisingly diverse, housing giant pacific octopus and their babies, five species of sea star, urchins, crabs, sea slugs, and anemones. Laura, an obsessive and brilliant underwater videographer, told me that she first noticed stars falling off the pilings in late October. She had been worried and went back repeatedly to film the underwater horror that was unfolding.

Laura does a lot of diving in the dark at night. A few weeks earlier she had recorded nighttime video footage that she directed me to watch on my laptop. I stared at the screen as hundreds of ochre, mottled, and sunflower stars peeled off the pilings, arm by arm as tube feet lost their grip, the scene ghoulishly lit by Laura's video lights. Some stars were so far gone that their bodies ripped away, leaving only an arm or two hanging, organs spilling out. Underneath, the pilings were surrounded by hundreds of decomposing stars slowly turning into a white bacterial mat. I told Laura I would come help and headed for the airport. It wasn't the first time that I had dropped everything to jump on a plane to document a disease emergency. The scale and pace of starfish mortality seemed ominous and important to see firsthand. Like a crime scene, the site of a massive wildlife die-off contains hundreds of critical details to notice and record. To begin the process of understanding this event, to fit together the puzzle pieces, I needed to see the scene for myself. How many stars were on a beach? How many stars were dead? What sizes were they? How many were dying? Were the sick ones grouped together? Were the sick stars grouped on a warm area of the beach or near pollution or freshwater stresses? Was the beach a rocky ledge? Was it composed of rocky cobble, gravel, or sand? Indeed, it would soon be apparent that we were experiencing the largest epidemic in marine wildlife history, one that demonstrated how frighteningly fast an outbreak can spread and virtually eliminate an entire chunk of ocean biodiversity.

. . .

Our oceans and the life forms they support are under siege, threatened by a formidable collection of forces that cause both sudden mass mortalities and a slow degradation of biodiversity. The top threats are the warming and acidification that accompany climate change, over-fishing, pollution from human activities on land, nearshore dredging, and oil extraction. Faced with such a ponderous list, it is hard to prioritize. As a marine ecologist specializing in disease, I worry most about the threat posed by microbes, because in oceans beset by all these stresses, microscopic disease-causing organisms can gain the upper hand, cause death on a massive scale, and thereby bring about rapid, wide-scale ecological change.

Microbes are scary in part because they are changeable and not under our control. Pathogenic organisms in the microbe category—viruses, bacteria, fungi, protozoans, and other disease-causing agents that don't fit neatly into these groups—are constantly evolving, their genetic codes often changing rapidly and staying one step ahead of their hosts' defenses. Think about one of the deadliest of human diseases, the Ebola virus, which causes fever, severe headache, vomiting, diarrhea, and hemorrhagic bleeding in its victims. Available evidence indicates that the virus has existed in bats in Africa for a long time, occasionally jumping to human beings but never breaking out beyond Africa. Then in 2013 a horrific epidemic of Ebola virus started in Guinea, Liberia, and Sierra Leone, spreading faster and further in Africa than previous outbreaks. Declared a Health Emergency of Special Concern in August 2014, it ultimately killed over 11,000 people and reached Europe and North America. Why was this outbreak so much bigger than earlier ones? Scientists aren't sure, but one hypothesis, backed up by intensive study of the changing viral genome during the epidemic done by a team led by Daniel Park of Harvard's Broad Institute, is that a key mutation allowed it to become more transmissible among humans. Because viruses have a very short life span and so many new virus particles are produced in the body of a single host, there is ample opportunity for such mutations to occur.

Microbes are dangerous too, because many can attack and infect more than one species or spontaneously develop that ability through a favorable genetic mutation. Pathogens that have a wide host range, called multi-host pathogens, tend to be deadly for at least some of the species they can infect. By deadly I mean they can kill every individual within a susceptible species, even driving them to extinction, while persisting in a more resistant host species. Contagious individuals in a resistant species can keep exposing healthy individuals in a susceptible species until the susceptible species is wiped out. In these situations, we say that the pathogen has a reservoir in the resistant hosts—a nice comfortable hideout from which to spread. The starfish epidemic of 2013–14 was caused by a multi-host pathogen. It affected almost all the major species of the entire group we call starfish.

Those of us who study non-human diseases are very concerned that mass mortality caused by multi-host pathogens is becoming a common and recurrent event, threatening biodiversity in both terrestrial and marine environments. The frogs of the world's rainforests are a high-profile example. Many rainforest frog species have been not just devastated but eradicated, at locations around the planet, by a skin-attacking fungus, *Batrachochytrium dendrobatidis,* called chytrid for short. This lethal fungus grows in the skin of the frogs, spreading its killing tendrils and exploding cells from the point of entry throughout the frog's skin. It can kill some species within days and is wildly contagious. In others, it lingers longer, so the infected frog continues to shed infectious spores from its skin into nearby streams. In 2015, using data from museum databases and field counts, John Alroy estimated the epidemic has caused the extinction of more than two hundred species of frogs. The chytrid epidemic has happened so quickly and in such remote tropical locations that the exact number of extinctions is still unknown, but we do know that many species are forever gone from our planet.

Lyme disease, carried by white-tailed deer, is also caused by a multi-host pathogen. As reservoir hosts, white-tailed deer can carry the Lyme disease bacterium (*Borrelia burgdorferi*) for many years before transmitting it back to ticks, which in turn pass it on to other species like humans. Rabies and Ebola are other examples of pathogens that have wildlife reservoirs and can infect humans.

Pathogenic agents with the potential to infect a suite of related species are bad enough in terrestrial environments. Put them in the ocean and you've got big trouble, since there is no way to get them out. Outbreaks in our oceans are different than the epidemics we face on land, in part because the oceans are a microbial frontier. Oceans harbor a far more complex mix of bacteria and viruses than exist on land, and our knowledge gaps about the biology of these diverse ocean microorganisms are much larger. In addition, to many of us, the oceans are out of sight and out of mind—we can't see a new outbreak beginning and act in time to collect the needed data or change the outcome.

To make matters worse, we have created a perfect storm of outbreak conditions in the oceans. Aquaculture and human sewage introduce new pathogens and fertilize existing ones—and it turns out that salt water is a hospitable environment for many pathogens that typically infect animals on land but can't live in air. Shipping spreads pathogens around the globe. Pollution and climate change weaken organisms' immune systems and thus their ability to fight new threats. The warming of surface waters from climate change makes conditions more conducive to pathogen survival over a wider area. Given the many ways we have mistreated our oceans and made conditions friendlier to microbes, it is no wonder that we are beginning to see more outbreaks of marine diseases.

The great starfish epidemic that opened this chapter—sea star wasting disease—is only the most recent of the ocean plagues. In the mid-1980s, the world saw its first case of a disease threatening an entire ecosystem when an infection decimated the two main reef-forming corals in the Caribbean. Near the end of that decade, several species of abalone along the Pacific coast of the United States began to sicken and die, victims of a slow-rolling outbreak caused by what turned out to be an unusual kind of bacterium. In 1984, a disease struck farmed Atlantic salmon in Norway, and subsequent outbreaks, caused by the same virus, have occurred sporadically in farmed salmon all over the world. In the 1990s, a disease caused by a fungus began killing coral species that hadn't been affected in the earlier bacterial plague.

Scientists have tracked and investigated these outbreaks, learning a great deal about their causes, courses, and consequences. I have been fortunate enough to be among them, and have led several teams of scientists to track outbreaks and consider solutions. It started with an international team that originated when the World Bank granted us funding to study the destabilizing ecological effect of coral disease outbreaks and to provide training in tropical areas such as Mexico, West Africa, and Indonesia. More recently, our team has expanded to include a focus on the health of temperate waters and to number more than forty scientists. With funding from the National Science Foundation, our Research Coordination Network for Ocean Health focuses on training the next generation of scientific leaders and developing surveillance of new infectious disease outbreaks to detect how climate change fuels these outbreaks in the ocean. There is still a great deal we don't know, but we have learned enough to be fearful about what could happen and hopeful that together, we can find ways to keep our oceans healthy.

·   ·   ·

When I'm at home on San Juan Island, enchanted by the natural beauty surrounding me, it's difficult for me to keep in mind the many threats facing the oceans and their bountiful life. The island, where I live part of the year, is one of several in a group in the Salish Sea, north of Puget Sound and Seattle. By all appearances, it's a marine paradise where people coexist harmoniously with ocean life. On land, the hills are clothed with forests of towering cedar and Douglas-fir trees mixed with red-barked madrona, and rolling farmlands fill in the spaces between the patches of pristine wilderness. The waters around the islands are clean, deep, and cold, part of a vast network of coastal waterways that stretch from Puget Sound far up the Strait of Georgia to Desolation Sound on northern Vancouver Island, and they encompass the richest coastal sea in the United States and Canada.

Our waters have, for some groups of marine organisms, the highest biodiversity in the world. The Salish Sea is packed with five species of salmon, jumping so thick some days during the summer run that it feels like they could leap into your boat. It's a good thing there are lots of them, because they have supported an impressive pod of southern resident killer whales, or orcas, each of which might eat three hundred pounds of salmon per day. From the deck of my house, I can look across our forest-ringed bay to the crossing of two waterways, Haro Strait and the Strait of Juan de Fuca, and past that to the shores of both Canada and the Olympic Peninsula. Best of all, our shore is on the daily run of that pod of southern resident killer whales. Many days in the summer, we see water spouts and the tall black fins of breaching whales as they pass by. I usually hear them first, the great exhalation that echoes for miles across still water. I am so well-tuned to that combination of sound and vibration that I stop whatever I am doing when the faintest orca echo reaches me, much like a dog would. Sometimes they wake me in the early mornings with their breathing and the breaching that ends in a thunder-clap of a splash.

The environment seems healthy and pristine. Most of the creatures in the long food chains from eelgrass to whales are thriving compared to those in more impacted waters. But under the surface, all is not well—not even here. In the aftermath of the starfish epidemic, sunflower stars are virtually absent, and other stars are much less abundant than they once were. The disappearance of the stars has had major ecological consequences. And the salmon—food for the orcas and humans—are in decline, with pathogenic microbes playing a role. The southern resident orcas are classified as endangered, with the declining salmon population a principal cause.

The mismatch between the outwardly healthy appearance of my home waters and what I know to be the underlying threats to their health helps me understand why it is that many people I talk to don't readily appreciate the seriousness of the situation. As long as seascapes remain scenic and we can still buy salmon at the grocery store, there seems to be no cause for alarm about the condition of the ocean. Why should we be concerned about disease outbreaks when each affects only a tiny proportion of ocean life? The answer lies, of course, in the words of the ecologists: everything in the ocean is connected. Since humans depend on the oceans for much of our food—worldwide, humans derive about 15% of their animal protein from the ocean—the interdependence of marine species matters. Disease outbreaks in non-food species can affect species we harvest in ways that we may not be able to predict and in ways that are painfully obvious. Corals are a case in point: we can't eat coral polyps, but the destruction of coral reefs in the tropics, due to a multitude of factors that prominently include disease, is having serious impacts on the many fish species that use reefs as nurseries and habitats and on which millions of humans depend for much of their food. Further, the species we do harvest for food are particularly vulnerable themselves, often precisely *because* we eat them and are thus likely to raise them in aquaculture, where pathogenic microbes can thrive.

Looking beyond our own self-interest as consumers of food from the ocean, there are other reasons to worry about the spread and multiplication of marine diseases. The oceans support much of the planet's biodiversity. Of the seventy phyla into which E. O. Wilson categorizes all life, from bacteria to vertebrates, twenty are exclusively marine; they occur in the ocean and nowhere else. While there are more total species on land, due to the huge number of terrestrial insect species, the oceans house the greatest diversity of life forms, including unique types of body plans, like the radial, tube-feet-equipped design of the starfish phylum, the echinoderms. The incredible diversity of underwater life, a product of billions of years of evolution, is a treasure endangered by the injuries we inflict on our oceans.

Because of the interconnectedness of life forms, biodiversity loss in the oceans is not just a matter of individual species being picked off by diseases or other killers. The loss of just a couple of reef-building corals in one locality, for example, can have effects that reverberate through the entire reef ecosystem, causing the loss of hundreds if not thousands of species dependent on the special environment of the reef. I learned this lesson early in my career when I watched disease turn coral reefs into rubble. With the decay of the physical habitat, the ecosystem lost its internal regulation and many species declined or disappeared. As coral reefs around the globe—sites of the highest marine biodiversity on our planet—continue to be imperiled by the deadly combination of ocean warming, sea-level rise, pollution, and opportunistic infection, so too are their rich assemblages of species. Kent Carpenter, with the International Union for Conservation of Nature, led a team of scientists from around the world who estimated in 2008 that fully one-third of coral species are now considered in danger of extinction and entire reef systems are at risk. The numbers have become even more dire following the extreme heat waves of 2015 and 2016, which decimated coral reefs around the world.

As the example of coral makes clear, marine disease is not the sole cause (or even, for many taxonomic groups, the major cause) of the loss of biodiversity in the ocean, but it is emerging as a causal factor that works in concert with the others. Many scientists have tended to view disease as a "canary in the coal mine," a sentinel of change in oceans suffering from the cumulative impacts of human activity. As an ecologist, I think now that infectious disease is more than just an indicator; it is itself the new major change agent in the ocean. A warmer, more polluted ocean is turning out to be a sicker ocean, and infectious disease is often what's felling the species that call the ocean home.

There is no reason to believe that the outbreaks I describe in this book are the end of the story. The microbes, ever ready to exploit the vulnerabilities we create for them, will spring on us new surprises. But science is not without its weapons. By studying the pathogen-host interactions at the root of disease outbreaks, we are learning how we might gain the advantage in controlling some diseases and devise policies that can prevent and limit the spread of future outbreaks.

CHAPTER TWO

Coral Outbreaks

A Global Threat to Marine Biodiversity

Through the window of my mask I see a wall of coral,
its surface a living kaleidoscope of lilac flecks,
splashes of gold, reddish streaks and yellows, all
tinged by the familiar transparent blue of the sea.

Jacques Yves Cousteau

We grabbed cameras and underwater slates, radioed a dive plan
to the surface team, and entered Hydrolab's porch-like entrance,
where our scuba gear hung on racks. The chamber, like the rest
of the underwater habitat, was filled with air we could breathe,
and its pressure was the only thing that kept the water from rush-
ing in. Sitting at the edge of the water, we pulled on our masks,
snorkels, and wetsuits, entered the water, donned our scuba gear,
and began the swim to our worksite in Salt River Canyon,
St. Croix. It was the beginning of a week-long research expedi-
tion in Hydrolab, an underwater laboratory, run as part of the
Man and the Sea program of the National Oceanic and Atmos-
pheric Administration (NOAA). Salt River Canyon in St. Croix
in the US Virgin Islands is a vast, coral-covered underwater

canyon that drops from its top edge at fifty feet below the surface to a depth of over three hundred feet. On that day in 1983, sunlight streamed from the surface through clear water, and huge plating corals covered the cliff faces as far as we could see. Among the corals were red, purple, and yellow sponges and purple waving sea fans and the occasional giant barrel sponge. It was a happy place with a color wheel of bright fish—tiny yellow and red wrasses, blue tangs, green and blue parrot fish, a pair of red and green cat-sized angelfish, and many others.

Scheduled to be at depths greater than fifty feet for five continuous days, we were doing what's known as saturation diving. Since we were breathing mixed-gas air under pressure, the nitrogen bubbles saturated in our blood and body tissues. As long as we stayed at depth, the nitrogen bubbles stayed in our blood and tissues and were not a problem. However, if we surfaced quickly, the pressure on the bubbles would be removed and they'd get dangerously big, like those that suddenly appear when you open a can of pressurized soda. It would be lethal for us to surface from this intensity of saturation diving without a slow thirty-six-hour decompression to allow the gas to fizz slowly and harmlessly out of our blood and body tissues.

To support our work, Hydrolab was well staffed with support divers who answered our radio calls from the reef on a twenty-four-hour basis, brought us full scuba tanks when we worked, whether it was day or night, and delivered our hot dinners in a pressurized container. We needed surface support because once the week-long mission began, we could not surface without that thirty-six-hour decompression period, no matter what happened—big storms, power outages, or sickness.

I was a graduate student, invited along by Tom Suchanek, then a professor at the University of California at Davis, to help with his project. The day before, Tom and I and two other male researchers had left the surface support boat to move into our home for five days on the ocean bottom. As we descended on scuba, the water was clear enough that we could look down and see Hydrolab, installed beside a thriving coral reef on the bottom fifty feet below. Hydrolab was modest in size and simple in design; it didn't look like much more than a huge, thirty-foot-high pipe with a porch and one big picture window. Inside, it was spartan, with three living areas: an entry porch, a main control room packed with instruments along one wall and a small table and bench for six, and a bunk room with tiny, narrow bunks but a huge, round window to the reef at the end. There was no indoor bathroom. Answering the "call of nature" in the middle of the night meant radioing to the surface, going on a solo underwater swim in the dark, and then confirming your safe return via another radio call. Conveniently, for these brief forays near the habitat we had an underwater hookah, a regulator with a very long airline. It was both slightly spooky and exciting to be out on the reef alone at night, and not so easy to settle back into sleep after a brisk swim at 3:00 am.

Our work on the bottom of the ocean was to document the competition for space between the corals and the bright sponges that were creeping out from under the corals and trying to overgrow them. A living coral reef is built and maintained by coral but is home to a diversity of other species such as sponges. Corals and sponges may seem like passive creatures, but they engage in brutal fights to the death to win space on the reef. When approached by a fast-growing sponge or another coral, a coral colony can rapidly grow new, deadly tentacles loaded with stinging cells to fight off the intruder. In the slow world of a sessile animal like a coral, such "rapid" growth might be on the order of three weeks. While the sponges are beautiful and important components of the system's biodiversity, the foundation and life-blood of the reef are the reef-building corals. Reef-building corals differ from soft corals and sea fans in that they lay down a massive concrete-like skeleton that creates the vast, intricate construction that is a coral reef.

On that first dive of our mission, the four of us followed an underwater rope trail from the Hydrolab to the edge of the drop-off reef. We reached the edge near our work site in about fifteen minutes, sat on a bench filled with fresh twin-eighty diving tanks, and changed into new working tanks. A pesky four-inch-long damselfish had set up its territory near the tank rack and repeatedly charged at our face masks and rather ineffectually snapped at my legs. The attack on the men's legs was more successful, since the wily fish figured out that yanking on leg hair made a diver jump. We shook our heads, laughing at the annoying, audacious fish and looked over the steep edge at the vast reef that stretched over 100 feet below us, covered with car-sized healthy brown and green plating corals, slowly swaying purple sea fans, multi-hued sponges, and patrolling fish. The deep reef below 60 feet feels very different from the frenetic pace of the shallower reef. The colors are more muted, there are fewer small fish darting, and it feels quiet and still and slow compared to the bright, fast action on the shallow reef. The corals are different shapes, with delicate, flat plating corals instead of sharp branching corals dominating. We descended to 120 feet and set to work carefully laying out our transect lines on the fragile deep reef and taking photographs.

One part of our research was novel and exciting: we were documenting the chemicals, brand new to science, being made by corals. We thought these chemicals might help corals keep other organisms, like the sponges, from overgrowing them. But we suspected that corals might also use them to fight their predators and pathogenic microorganisms. Understanding the details of how chemicals like these mediate biological relationships in nature would turn into my life-long work. Scientists call these chemicals biologically active because they are potent in disarming biological systems, with destructive properties and almost god-like effects on other biological entities. They can inhibit the growth of or kill the toughest bacteria, viruses, and fungi. They can make fish sneeze, vomit, swim away, or die. Some have effects on living systems that are more subtle; anti-inflammatory chemicals made by soft corals can reduce redness and inflammation in human skin, and we can only wonder what they do for the corals. It was the study of these chemicals that eventually lured me into researching the warfare between disease-causing microorganisms and the corals' natural defenses.

The adventure grew day by day. We would work a full eight-hour day underwater and then come back to our dry underwater home cold and exhausted but giddy with nitrogen narcosis and excitement over our adventures. The stupidest jokes could double us over for minutes at a time. Then, warm and dry at day's end, we would watch through our giant window as the day fish swapped out with the night fish. As the light faded, schools of giant rainbow parrot fish would journey off the reef to deeper overnight sites. Once the dark descended, clam-shaped ostracods the size of sesame seeds would begin their nightly ballet. We could look out our picture window and see their dance of light. Unlike a firefly, which holds an internal light in its abdomen, a male ostracod releases tiny blips of bioluminescent fluid that stay in place for five to ten minutes. As the animal swims from the reef upward, it creates a column of lighted dots that interested females can follow. Each species has its own pattern.

A common pattern is a vertical column of ten tiny dots of light, each spaced six inches apart. The many busy male ostracods replicated their traces of light numerous times over, leaving hundreds of tiny beacons hanging over the night reef, a tapestry of lights that lit up the water as far as we could see.

Whether living in an underwater laboratory or forging through deep jungles, field biologists live for this mix of making new discoveries in the wild and being together on the adventure. The four of us loved every minute of our week underwater. We worked hard and bonded in that special way that comes with sharing an intense experience. To this day, thirty years older and gray-haired, the four of us trade secret hand signals when we run into each other at scientific meetings. Cracking goofy jokes about Roger, the imaginary aquanaut we blamed for all our problems, can still double us over laughing.

The Hydrolab mission was my early exposure to coral reefs. As a side project during that mission I studied how a brightly spotted snail, the flamingo tongue snail, attacked and ate sea fan corals. Sea fans make very toxic chemicals and I was puzzled as to how this beautiful snail could eat such poisonous prey. It turns out they are able to alter the chemicals and thereby detoxify them. On land, monarch butterfly caterpillars perform the same kind of detoxification when they eat the very toxic milkweed.

．　　．　　．

After the Hydrolab mission I finished my PhD and was hired as a professor of marine invertebrate biology at Cornell University. I recruited a research team of PhD students and Cornell undergraduates and together we pursued the puzzle of what gorgonian soft corals were doing with the biologically active, often toxic chemicals they produced. I was interested in the functions of these chemicals in nature, but there is also an applied side to this research. Could some of these chemicals fight diseases of humans? To that end, we collaborated with some chemists working with pharmaceutical companies to investigate the potential to develop human drugs from the sea. Gorgonian corals were identified by ecologists and chemists alike as a group with high potential for important, biologically very active chemicals. Gorgonians are the sea whips and sea fans in the group of corals called octocorals, so called because each polyp has a ring of eight tentacles rather than the usual set of six. Like most corals, sea fans and sea whips are colonial organisms made up of many individual genetically identical polyps, so one "individual" sea fan or sea whip is properly termed a colony. One of their chemicals, made by the sea plume *Pseudopterogorgia elisabethae,* is pseudopterosin, which has potent anti-inflammatory properties and almost made it all the way through years of clinical trials as an anti-arthritis drug. Other examples are the super-potent briaranes and asbestinanes made by the gorgonian *Briareum asbestinum.* Briaranes are wonders of nature, large molecules with over twenty carbons and nasty elements like bromine and chlorine that make them deadly to fish and dangerous for humans. One of my PhD students at the time developed a serious sensitivity to this compound and couldn't be in the same room with even dried samples of *Briareum asbestinum.* We began our work by investigating whether these chemicals helped defend the corals against snails and predatory fish. Fish would not get near them, but we also noted that these chemicals scored very high in initial tests against microorganisms like bacteria and fungi. These tests were designed to screen for new naturally occurring antibiotics that might be useful in human medicine. The success of these chemicals in reducing pathogenic bacteria in the lab made us wonder if another of the chemicals' functions was to deter infections in nature. We got our chance to explore this question in 1996, following reports from all over the Caribbean of a mass die-off of sea fans.

I first heard about these declines when one of my colleagues, Jim Porter, a professor at the University of Georgia, called me in Ithaca one day in 1994. He'd heard at a meeting that sea fans were dying at many locations around the Caribbean. "If you need samples," he told me, "this is the time to get them; your populations are disappearing quickly." In the following months, I communicated by phone and email with researchers who studied these animals and we all became unnerved by the growing reports of mortality. Given how widespread the die-off was, I was worried about the impact it could have on sea fan populations. We launched a study to document the ecology of an underwater epidemic and monitor the defensive function of the coral's chemicals in nature.

My career was built on studying inducible defenses in invertebrates—those spines, tentacles, or chemicals activated to new, fighting levels by a predator or pathogen. How could I test the hypothesis that the sea fan chemicals we had been studying were actively fighting this new threat? What *was* this new threat? First I had to figure out what organism was killing the sea fans. A microbiologist named Garriet Smith, from the University of South Carolina, had been working on the sea fan epidemic in the Bahamas. He had published a paper in 1996 with compelling evidence pointing to an unlikely culprit: the fungus *Aspergillus sydowii*, which normally lives on land, in the soil. He and his students had done careful work to come to this conclusion. They were able to isolate cultures of the fungus from sick colonies of sea fans and show that injecting a mix of spores and tissue from these cultures into healthy sea fans would make them sick. Interest mounted with the realization that corals and humans might be infected by a common pathogen. A closely related species, *Aspergillus fumigatus*, causes a fatal respiratory disease in immune-compromised humans. But how was a fungus from land infecting sea fans?

There were two hypotheses about how infective particles from the soil could reach the sea fans. A scientist working for the U.S. Geological Survey, Eugene Shinn, had noticed that huge clouds of dust from the vast deserts of Saharan Africa were being blown into the upper atmosphere and transported across the ocean. There were days when the dust was thick enough to cloud the skies in the Bahamas and parts of the Caribbean and settle the distinctive red soil in thick deposits in the Bahamas. This African dust had a very high iron content, and Shinn hypothesized that spores of the soil fungus were traveling with a built-in fertilizer, iron, that helped the fungus grow when it hit seawater. Iron is often a limiting nutrient for microbes growing in low-nutrient seawater. Although one of my students typed the molecular signature of dust samples one year and did find *Aspergillus* spores in the dust, it was not an exact species match with the pathogen affecting the sea fans, so this remained untested.

An alternative hypothesis was that the fungus was more local and had been introduced to the ocean in coastal runoff. Prior to the 1994 outbreak, huge rains had drenched South America and breached the banks of the Amazon. The runoff plume from the Amazon, one of the largest rivers in the world, has a startling reach into the Caribbean. Following large storms, the Amazon freshwater plume can spread northwest through the Antilles and almost to Puerto Rico. Ivan Nagelkerken of the University of Adelaide described in 1997 how Amazon waters were detected as far north as Trinidad a year before the outbreak, in 1993.

I had to see firsthand what the pattern of mortality looked like both within colonies and across the reef before I could design experiments to learn more about whether the sea fans could fight the infection using their defensive chemicals. At the same time, I got caught up in the mystery of how these infections were starting and where they were coming from. This marked the beginning of my entry into the world of disease ecology. I needed to see how the infections progressed within a colony—did they start at the older base of a colony or at its younger top edges? In other words, was old tissue more susceptible than young? Was there a consistent pattern, for example of spread from base to tip, repeated across many fans? What were the patterns within the population on the reef—did the largest old fans die first or the young babies? Or was there no consistent pattern, with lesions localized around bites from snails and fishes, indicating the role of a vector? Which underwater locations were the most vulnerable—shallow ones, near land, or those in deeper water? I realized I needed help to answer these questions.

When I look back on my career, I can see that picking talented and trustworthy collaborators has been a foundation for success. The good move has always been to work with people who are smarter than I am or have skills I lack. As an ecologist, I had limited training in microbiology and no expertise that would help me verify that *Aspergillus* was the infective agent. My next move was crucial. I spent the better part of a day figuring out how to word an email to Garriet Smith, having no idea how it would be received. Would he be territorial and not appreciate having another scientist, especially an ecologist, barge into his study? It was one of the happy moments in my career to get an enthusiastic and encouraging reply within about seven minutes. "I really need an ecologist's help," wrote Garriet. "Could we work together at a field site?" I was lucky: Garriet Smith, distinguished microbiology professor at the University of South Carolina, is one of a rare breed of scientist who has the skills to isolate and grow microorganisms in the lab and type them with new molecular methods. And he had an interest in studying their interactions with other organisms in nature. Serendipitously, we were both doing some of our research on the tiny, rather remote island of San Salvador in the Bahamas, and Garriet had identified this as an active outbreak site, so I planned to join him for a joint project there. Thus began my experience with infectious disease.

•   •   •

I disembarked our small charter plane on San Salvador in January 1996 with a newly hired postdoc, Dr. Kiho Kim, who had recently completed a PhD research project on coral population biology in the remote San Blas Islands of Panama. In addition to being a fabulous cook, Kiho is now a professor at American University and director of AU's Center for Teaching Research and Learning. Immediately hit with overbearing heat and humidity, we crossed the tarmac, hoping our ride to the lab was on time. We would only be there for a week and wanted to get right into the water. We heard a rumble and I looked up with a nod, pleased to see a huge 1920s-vintage blue truck lumbering down the road to the airport, the back loaded with fifteen students who were here to study the reefs for a class. We hauled our gear into the truck: three duffels loaded with scuba gear, three metal storage boxes with our lab equipment and supplies, a cooler for shipping samples back, and one folded-up Zodiac boat with an outboard motor.

This tiny island in the Caribbean region had been the site of my lab's field research since 1989. The Gerace Marine Lab, our base on the island, was a very basically equipped field station. But its immediate proximity to a stunning coral reef and pristine waters was an undeniable virtue. I knew very well that the white sand beaches and azure waters we saw flying in were deceptive in their suggestions of paradise. In the evening, we would be prey to vicious hordes of mosquitoes and, even worse, no-see-ums that would slip through the very basic screens of our non-air-conditioned housing. Some of my students and my husband grew to hate this place, with its burning sun and hungry bugs, but I liked its remote simplicity and natural beauty. About six hundred people lived on San Salvador, and the waters were pristine and clear, the reefs largely intact. Indeed, they were so intact that the water was full of all kinds of fish. On dives, we were frequently visited by large predators like barracuda. On several deep dives we had surfaced to our safety stop at fifteen feet only to be confronted by circling hammerhead sharks, close enough to feel them focusing their unsettling, eerie stalked eyes on us.

We wanted to get a look at the sick sea fans as soon as possible. There was just enough time in the afternoon, after arriving at the lab and grabbing lunch and unloading our gear, to get to one of the nearer sites. We pulled together our snorkel gear, hoping the sea fan populations would be shallow enough for us to reach them just by snorkeling. The class was headed our way so we hitched a ride with the students in the back of the ancient truck. We had left Ithaca in the midst of the coldest, darkest, and shortest days of winter, so the sun felt especially good on our faces as we raced along one-lane roads with palm trees and pristine white beaches flashing by.

We arrived at a site called Telephone Reef, where Garriet had said we might find an active outbreak. The driver dropped us off at a deserted beach, promising to return in two hours. It had been a busy three weeks preparing for this trip, followed by many hours of travel and meager sleep. It felt good to sit in the fine white sand on a deserted beach, clear blue water and reef stretched in front of us. I breathed in the sun, wind, and beach smell. Kiho sprawled out on the warm sand, eyes closed. We could have easily stayed on that beach, but we looked toward the lowering tropical sun and headed down the beach for the reef.

I cleared my mask and pushed through breaking waves in the surge zone and dropped underwater. Once past the waves, it was a clear, calm day underwater. I glided over a sandy bottom patterned with ripples; brown ridges alternating with beige valleys like the dunes of some vast desert seen from the air. We approached the reef, swimming over larger and larger coral heads with tiny fish darting in and out of their complex underwater architecture. The site was like a garden. Hundreds and hundreds of tall purple-lavender and bright lemon-yellow sea fans waved back and forth in the five feet of water, lit by fading late afternoon light. I searched for signs of damage, but the fans all looked to be in good health. Then the full expanse of the deeper reef opened out in front of me, coral heads stretching as far as I could see. Brown and green *Orbicella* mounding corals were mixed with some gorgeous old dark-green *Diploria labyrinthiformis* and *Colpophyllia natans* brain corals.

We kicked hard to reach the farther side of the reef, where we started to see purple sea fans with gaping holes surrounded by dark purple edges. Kiho pointed at them and I nodded. This is what we had come to see, and it looked bad. Some fans had multiple lesions, big holes running up along the central veins of the colony and expanding to create dead holes, with the entire fan mesh eaten away. My brain was in overdrive processing what we were seeing. We swam over to a section of reef that looked like a sea fan graveyard. Fan after fan had been stripped of all living tissue and reduced to skeleton—stumps with bare branches sticking up eerily. These fans had died so recently that their skeletons had yet to be grown over by algae.

It was a grim sight, but it meant our trip was necessary. I started going through my mental checklist: size and condition of lesions, size of affected colonies, areal extent of the mortality, appearance of the lesions, distribution of sick, healthy, and dead fans on the reef. Looking closely at the lesions, I could see that this infection was serious. There was a zone of new tissue death at the edges of the lesions, grading into complete loss of living tissue in the centers. There was no algal colonization, so the dead area was new and fresh. Where the dying tissue joined the living tissue, the living tissue was swollen and darkened, creating pronounced purple halos (see figure 2). This suggested to me an inflammatory response, possibly an active defense by the sea fans. This was what I wanted to study: the chemicals that were pumping inside those purple halos. Those chemicals could be the secret to success of the most primitive immune system on the planet.

We swam across the reef for over an hour, taking in every detail of the fans and lesions, talking about how to take the right kinds of samples and data. Finally, the setting sun and fear of large nocturnal predators drove us out of the water. The truck was waiting, the students impatient to get back for their cold showers and dinner.

The next morning, we headed to the airport to meet Garriet. The small plane taxied in on a clear day and folks filed off. We didn't know what Garriet looked like but spotted a middle-aged guy with dark hair and big glasses, dressed conservatively in khakis. He looked like a microbiologist. "That must be him," I

Figure 2. Lesions in a sea fan caused by *Aspergillus sydowii*. The dark purple halo around each lesion indicates where the coral's immune system is fighting back with antifungal chemicals. Photo by Ernesto Weil.

nudged Kiho. He nodded. Neither of us noticed the smiling, white-bearded guy in shorts and t-shirt headed our way. "Dr. Harvell?" I turned in surprise and replied, "Dr. Smith?" "It's Garriet—I'm pleased to meet you."

I introduced Kiho and we helped Garriet with his gear and headed for the truck. Garriet had often brought groups of students on field courses to San Salvador, so during the drive we chatted comfortably on the neutral topic of the joys and trials of working in this place. Then the three of us sat in the dining room at the Gerace lab to eat a late breakfast and make a plan. Kiho and I had many questions about how Garriet had first come to suspect *Aspergillus* as the pathogen and the methods he had used to isolate it from the sick sea fans. We finished eating and then walked over to the lab, where he unpacked one of his boxes and showed us a petri dish full of live *Aspergillus* he had earlier isolated from a San Salvador sea fan and brought along.

This was the moment I was hooked. Not only was Garriet a very knowledgeable microbiologist, but we had similar values about what we thought was important and interesting. And he had what we needed the most: a cultureable pathogen that we could use to inoculate live sea fans and study how their immune system responded. So began our decades-long collaboration. Started in the shared need to solve the puzzles about this novel outbreak of sea fan mass mortality in 1996, it would expand as we sleuthed multiple later outbreaks in other coral species. Our collaboration was to be marked by many joint international trips, shared training of students, and solid friendship.

After we collected healthy sea fans and installed them in running seawater aquaria, Garriet showed us his idea for how to inoculate them with *Aspergillus* cultures he had isolated from these reefs on a previous trip. His initial method was to break up the culture growing in the petri dish to smaller hyphal fragments and spores, dilute these samples with seawater, and inject the mixture into target fans. He inoculated one group of target fans in several tanks and kept another un-inoculated group in separate tanks, as control fans. Quarantine facilities were not adequate at this lab back then, so even though the *Aspergillus* had originated on the San Salvador reef, we did not want to add any back. As a precaution against any infectious spillover, we bleached the seawater when it drained from our tanks. Each day after that we checked them. Five days later, lesions started at the sites of the injections and the controls stayed healthy. This experiment was a practical demonstration of Koch's postulates, the central tenets in infectious disease research. Koch's postulates describe what is needed to prove the identity of an infectious disease agent. First the agent must be isolated from a sick person or organism and put in pure culture, then it must infect a healthy individual, and finally it must be re-isolated from the newly infected individual. Fulfilling the requirements of Koch's postulates is a big challenge in the ocean, because most marine pathogens are not easy to culture. Even worse, sometimes infectious disease is caused by several microorganisms working together. We would end up working with the *Aspergillus*–sea fan pathosystem for a long time, in large part because *Aspergillus* is a cultureable agent and it was key to unlocking discoveries about the sea fan immune system.

Even as Garriet was inoculating the fans, my mind was racing ahead, thinking of possibilities for other kinds of methods we could adapt and maybe employ in field experiments. We spent much of the rest of our time in the lab talking, planning more ecologically relevant ways to run these experiments. Rather than injecting the fans, should we be releasing the culture slurry directly into seawater in quarantined tanks, a process closer to what might happen in nature? We also considered ways to apply *Aspergillus*-laden mesh, like Band-Aids, to the colony surface. These methods would be the key to studying how sea fans reacted to controlled infections and would result in a series of papers published over the next decade.

Our other starting point was to survey the health of the sea fans on the reef in this mounting outbreak. How quickly were their lesions growing, and how many were really dying? Kiho and I spent hours and hours underwater for the remainder of the San Salvador trip. A typical day for us was to get into the water early, before the waves mounted, to lay our transect lines and count and measure all sea fans along a seventy-five-foot transect.

For each of the hundreds of sea fans, varying in height from the babies at two inches to the venerable old fans at five feet high, we would assess health, count the number and size of lesions, and record height. Although we focused on tabulating data on our dive slates, we could also stop occasionally and feel the rhythm of the reef. We were followed by a flotilla of fish that were hoping we would dislodge a small crab or marine worm they could eat. We saw brightly striped sea slugs, spotted flamingo tongue snails, and huge conchs.

One day Kiho and I were working along a transect, one on each side, when we ran into a barracuda. It was huge by our standards, over four feet long and very deep bodied, the biggest we had seen. I hadn't noticed it, too busy with my head down counting. But Kiho reached over, nudged me and then pointed. Oh! I nodded in surprise at the size and how close it was. We both stopped and watched. The barracuda wasn't going anywhere. Kiho made a very polite "after you" gesture to me. This made me laugh, which loosened the seal around my face mask, allowing water to flood in. I stopped laughing and cleared my mask. Was he really serious? He wanted me to go first? Then he moved in behind me and I laughed again.

I decided to get it settled with the barracuda right away. I stood upright, raised my arms, banged loudly on my tank and made growling noises. Absolutely no response. The barracuda just looked at us more intensely. Then it made a threat display, clacking its jaws open and shut several times in succession. I imagined all the fish on the reef suddenly stopping to watch this drama. I was scared, but I didn't think it would attack us and I was determined to impress Kiho. So I made an even bigger forward lunge. Bingo—it worked. After one more angry mutter and jaw snap, the 'cuda moved off. I rubbed my hands together in an exaggerated way as if to say "took care of that, all in a day's work." At the time I thought Kiho was actually scared, but now I think he was being generous and giving his foolhardy adviser a chance to feel tough.

We didn't know it then, but we would work together for more than ten years, with Kiho leading the field program, running many belt transects and counting and measuring thousands of sea fans at sites from the Florida Keys to the Bahamas. We discovered that the epidemic slowly waned over ten years as susceptible hosts died and were replaced by resistant ones. The most interesting part of the story was what was going on in those purple halos, an active zone of sea fan chemical defense.

Kiho and I were swimming transects in the Florida Keys to study sea fan health when the big El Niño hit in the fall of 1998. On those transects, we noticed many of the corals turning white as the temperature mounted and the corals lost their symbiotic algae and pigments. A spectacular change happened when a bright purple soft coral, *Briareum asbestinum,* suddenly turned overnight, from dark purple to a whiter shade of pale. There were a lot of them on our transects and we had never seen *Briareum* bleach before, so we added another column to our data sheets and started counting the frequency of bleached colonies. What happened over the next week was unexpected. The skin on the bleached corals began to fall off in necrotic patches, followed by mortality. All told, an average of 68% of the bleached colonies died. We immediately dropped all our other underwater work to focus on finding out what was killing them.

The essential question was whether they had an infectious disease triggered by warming or were dying from heat stress. We had no laboratory to work in or supplies for complex experiments, but we had what we needed most: good field sites. We set up an underwater experiment to test if the sick corals could transmit an infectious agent. It was sort of like a gardener's plot. We clipped branches from sick corals and zip-tied them as grafts to nearby healthy corals. As a control, we zip-tied healthy branches of *Briareum* to healthy *Briareum*. The results were clear and statistically sound: the necrosis was transmitted from the sick to the healthy corals in half of those cases and in none of the controls. We had confirmed experimentally the transmission of an infectious agent that was killing the heat-stressed corals. We sent samples to Garriet Smith to see if he could identify a microorganism that could be causing the sickness. He isolated a cyanobacterium, *Scytonema,* that he posited as an infectious agent, although we couldn't rule out that more was involved. A year later, the sick colonies on that reef had died and disappeared and the surviving colonies had recovered, losing their necrotic lesions and regaining their color. If we had not been surveying those reefs in October, during the warming event, all we would have noticed was that there were half as many colonies remaining the following year. It would have been one more unrecorded disease outbreak.

. . .

Sea fans are in the taxonomic group Cnidaria, the phylum that includes anemones, corals, and jellyfishes. The group contains some of the most ancient animals on our planet, its members very different from us in having no backbone, only two cell layers in their bodies, and the simplest nervous system. The antiquity of the cnidarians made it biologically exciting to investigate how they defend against microorganisms in nature. We know that cnidarians, like most invertebrates, rely solely on what is called innate immunity, which comes primarily from the production of compounds with general efficacy against broad classes of pathogens. They lack the adaptive immune systems of humans and other vertebrates, which work in concert with innate systems and can react to, target, and remember specific strains of a pathogen. But we know surprisingly little about how the innate immune system works in this ancient animal lineage. In what ways is it different from our more complex immune system? Obviously it works well, since these animals live and usually thrive in a microbial soup of potentially harmful bacteria, viruses, and fungi, not to mention larger parasites.

By the time we expanded our sea fan research from San Salvador in the Bahamas and the Florida Keys to the Yucatan Peninsula in Mexico, we were three years into the outbreak. Millions of the most susceptible sea fans had already died. But then something changed. Studying the survivors, we found, to our surprise, that the lesions stopped growing and the fans started recovering. By 2002, sea fans were not melting away as they had done in 1996. It was of course great news for sea fans, but also kind of a setback for us, since we had expected to see the ripping-fast death and destruction we had observed early on in the Bahamas and then Florida.

We had already started to study the powerful immune system of the surviving sea fans and identified chemical components that could slow this deadly pathogen. Had the sea fans become immune to the fungus? We could watch the evidence of successful resistance in front of us—lesions where the fungi started to kill polyps, development of purple pigmented halos around the lesions, then the infection subsiding, then a return of coral growth. As we looked around the reefs, we saw more and more evidence of this resistance pattern—tattered fans with large holes and rips, but no new mortality and, in fact, new growth sprouting. The following year, the reef in Akumal, Mexico, was covered with sea fans that had disruptions in the patterns in their mesh, signs of old infections, but they were growing new healthy tissue. We wanted to figure out what had changed: the environment, the corals, or the fungus? We conducted lab experiments. The same cultures of *Aspergillus* were no longer causing lesions in our lab experiments. This led us to an intensive few years of studying the immune defenses of the sea fans.

A couple of years later, in 2003, I was puzzling over the latest results from experiments we had run. Earlier results were consistent with our hypothesis that corals make chemicals that stop fish from eating them. Not only did most fish not eat the sea fan corals themselves, they also avoided or spat out the artificial foods we put on the reef as taste tests. We were initially trying to understand if those chemicals had evolved to deter fish predation. But the story that all the biologically active and downright toxic chemicals in corals had evolved as a defense to keep fish from eating them just didn't fit comfortably in my head; our work with the *Aspergillus* fungus was showing that often, unseen microbes were also challenging the sea fan immune system. So now we were testing the chemicals against bacteria and fungi in petri dishes.

One afternoon, I heard footsteps coming down the hall toward my office and knew exactly who it was: Dr. Laura Mydlarz, then a postdoc in my lab. She gave a quick knock on my door and burst in, still wearing her lab coat. Tall, dark-haired, and intense, Laura is a chemist and an expert on natural chemicals made by animals like corals. "I think you'll want to see this—it's a little different," she commented in her typical understated way. "I just emailed you the results." With a nod, I turned to my computer and opened the Excel file she had sent, knowing she would not barge in this excited for no good reason. I scanned the data sheet, following down the column labeled ABA, for our antibacterial activity assay. The column reports the size of zones of bacteria that are inhibited by a particular chemical. It quantifies the bacterial killing potency of either a single chemical or a complex cocktail of several. It's a satisfying test to run because you see your results immediately and there is no arguing about what they mean: a large clear zone on the media beside the filter paper where bacteria do not grow means the chemical on the filter paper inhibits the bacteria. The chemicals leaching off the filter paper prevent the bacterial cells from growing and might even kill them. In Laura's results, the bacteria grew quickly right up against the untreated filter paper—our controls—and didn't grow next to the ones treated with our sea fan chemicals. These zones of bacterial inhibition were big, comparable in size to the zones we tested with a commercial antibiotic.

Then I looked down the columns for AFA, antifungal activity. Yikes, no wonder Laura was so excited! Fungi are wicked hard to kill. These extracts were preventing the growth of commercially bought cultures of *Aspergillus sydowii*, classified as a human pathogenic fungus, in a half-inch-diameter circle around the chemical disc. The results suggested that no bacterial or fungal species were able to grow in the presence of several of the chemicals from the sea fans. Not just one or two active chemicals, but multiple chemical fractions in our sea fan samples stopped bacteria and fungi from growing. Of course, we had expected the sea fan chemicals to show some antibiotic properties, which is why we were doing these experiments. But this chemical cocktail was protecting sea fans from both bacteria *and* fungi, and doing well against both. We now had good reason to believe that this antimicrobial chemical defense, one of the oldest natural defense systems on earth, is what makes life in the microbial soup of the oceans possible for sea fans. The chemical defenses that are a mainstay of life for sea fans are likely also essential for many other invertebrates in the oceans.

I looked up and met Laura's eyes. We both smiled and started talking at once about the same interpretation: the fish-deterring aspect of the chemical defenses in these corals is just a by-product of their toxicity—a bonus. The real driver of chemical diversity in these corals is more likely defense against pathogenic microorganisms. I sat back, relishing the satisfaction of an "aha!" moment. Our work on chemicals produced by soft corals—in collaboration with a natural products chemist at Scripps Institution of Oceanography—had been promising in that it showed that these chemicals deter or poison fish. However, I had always felt that we had only part of the story of why soft corals were so darn well defended by diverse chemicals with potent biological powers. Laura took the work one step further in a 2002 paper, showing that some of the chemicals could actually be induced to higher levels when exposed to the *Aspergillus* fungus.

The sea fan outbreak that had started in 1994 lasted about seven years and then slowly faded. By 2004, when we were fully geared up to study the epidemic, it had run its course. Studies of the decline of other epidemics reveal the action of three possible processes. Sometimes the hosts develop or evolve enough immunity to disarm the pathogen. Measles epidemics of humans wane on their own because the human immune system is adaptive: antibodies are induced, making each person less susceptible once they have been infected. Over time, the entire population becomes immune and the measles virus loses its hold. Sometimes the pathogen itself evolves to a genetically less destructive form. And sometimes changes in the environment cause epidemics to subside. In the end, it is always interactions within the

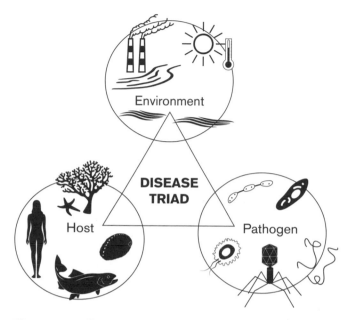

Figure 3. A traditional model of infectious disease causation known as the epidemiologic triad. The triad consists of an infectious pathogen, a host, and an environment in which host and pathogen are brought together. For some outbreaks, an environmental threshold, such as increased water temperature, must be passed for infections to start and an outbreak to occur.

disease triad—host, pathogen, and environment—that govern the dynamics of an outbreak.

Using the disease triad as our guide (see figure 3), we can identify the processes that govern how diseases take hold and expand into outbreaks, whether on land or in the sea. We can see the importance of knowing which microorganism fills that pathogen bubble and causes a die-off. We can see that it matters a lot whether our host bubble contains a human or other mammal with an adaptive immune system, or a starfish or coral governed by innate immunity. Finally, we can start to unpack the complicated ways that environment can drive every aspect of the dance between host and pathogen, often tipping the balance one way or the other. The rather hard part of doing the science is that in each outbreak, we have to investigate anew which component of the triad is responsible for the tipping point being reached. Is it a change in host, pathogen, or environment that destabilizes a balanced host–pathogen interaction and allows it to explode into an epidemic outbreak? It can be as simple as the introduction of a new, more virulent pathogen that leaps into a susceptible host, spreads rapidly, and begins an infection and killing spree. This is what happened with the Ebola outbreak; a new variant of the Ebola virus was transmitted from wild apes to a human. Or it can be that the environment has turned bad for the host and good for the pathogen, as is the case with many coral diseases and hosts that are stressed by warming and are newly susceptible to opportunistic pathogens that are fueled by warm temperatures.

Whether on land or in the sea, the main framework scientists use for thinking about whether a new disease will turn into an outbreak is the now classic mathematical model developed by Kermack and McKendrick. The model addresses how fast an outbreak will progress by separately accounting for how the disease is transmitted, how lethal the infectious agent is, and how strong the host's immune response is. These were named SIR compartment models by the famous mathematician Robert May, then of Oxford, and the epidemiologist Roy Anderson, of Imperial College, because the infection status of the host is characterized as susceptible, infected, or recovered. These models have been used most successfully to predict the time course of human disease outbreaks such as measles, but led by John Bruno, then a postdoctoral fellow in my lab at Cornell, and Steve Ellner of Cornell, we applied them to our coral disease outbreaks to estimate how long an outbreak would last and whether developing immunity in the coral host was the likely cause of the outbreak's end. The models allow a detailed simulation of the importance of different mechanisms, which can determine the final degree of outbreak.

In modeling the decline of the sea fan epidemic, we focused on changes in the model that describe the immunity of hosts. Vertebrates' adaptive immunity, based on the stunning ability of antibodies to track and remember viruses and kill ones that have previously caused infection, gives them many advantages in the fight against disease. In the SIR models, a virus fails to reinfect a vertebrate host that has recovered because the host has acquired immunity. When this process is replicated across an entire population, the acquired immunity prevails, the disease diminishes, and the epidemic fades out. What about marine invertebrate hosts that don't have the sophistication of adaptive immunity?

The components of innate immunity in our sea fans are relatively non-specific but extremely fast acting. In the innate immune response, there are inducible components like immune cells and antibacterial and antiviral chemicals that lead the charge and can also trigger the release of even more chemicals, thus scaling up the immune response. Many of these components stay activated after first contact with a microbe, providing longer-term protection akin to acquired immunity or immune memory. Scientists who study vertebrate immune systems tend to look down on invertebrates for their lack of acquired immunity, but my view is that the innate immune system, which has been battling microbes for millions of years in the sea, has unusual, under-appreciated powers for quick response and deserves more study.

My lab's studies over the past two decades have opened a window into corals' fast-acting innate immune system and their extremely aggressive immune cell response. For me, our most interesting findings have to do with what's going on in those purple halos, the active zones of sea fan resistance. Immune cells jump to the attack and proliferate around fungal invaders. They eat up some of the fungal strands and beat back others with anti-fungal chemicals, but they also get right to work building walls. They release tiny purple and black melanin granules that can wall off the fungi and stop them from penetrating tissue. Laura Mydlarz, who detected the huge antibiotic and antifungal activity, has made some amazing images of sea fan immune cells actively building these walls. Thanks to more work by people in my lab group, we now know much more about how the sea fan immune system works. When Colleen Burge joined my lab as a postdoc, we stopped using the old method of testing the activity of each chemical in a petri dish and began employing the newest tools in genetics, which allow us to study the activation of hundreds of possible immune genes in response to infections. The work on sea fan immune genes continues in my lab today, twenty years after our initial work with the fans, with the experiments of PhD student Allison Tracy.

In the case of sea fans, we were convinced from our field monitoring, experimental work, and modeling that the decline of the outbreak was caused by a rise in immune sea fans as susceptible hosts died and were replaced by resistant ones through a process of natural selection. We directly observed millions of sea fans die, over thousands at some particular sites. Our modeling showed that the high level of mortality we observed is more than enough to select for very potent resistance in the few survivors. When the resistant survivors reproduce, their new babies are also resistant. A new study by Jason Andras of Amherst College, published in 2017, provides genetic support for the hypothesis of newly resistant sea fans he sampled in the year 2003. Although we started our study with the outbreak and a focus on dying sea fans, we ended with new discoveries about sea fan resistance and a focus on those that survived the extreme outbreak.

. . .

The mass mortality of the sea fan outbreak that started in 1994 was just the beginning of an unfolding ecological disaster on coral reefs that continues to this day as the oceans warm. Corals are solar-powered animals that thrive in the nutrient-poor but sunny waters of the tropics through a symbiosis with tiny photosynthetic algal cells called zooxanthellae. Since all reef-building corals derive most of their energy from the algal cells living in their tissues, the vast underwater cathedrals that are coral reefs are built by solar power. Although this symbiosis is vital to the success of the reef, it is also the Achilles heel of the reef-building corals; the algal cells have a greater sensitivity to high temperature than the corals, and the symbiosis breaks apart during warming events, when the algal cells are expelled. This is called bleaching because the corals turn white when their colorful symbionts depart. The white is the underlying skeleton showing through the clear, transparent skin of the coral. Even a warming event that can seem insignificant, perhaps an increase of only 1–2 °F, can cause the corals to lose their symbionts, their main source of nutrition, and become stressed and often susceptible to opportunistic infections. The corals die from a combination of heat stress, starvation, and lethal infections.

White band disease was first noticed in the 1970s in thriving gardens of Caribbean staghorn corals (*Acropora cervicornis*). After the warm water conditions of the 1982–83 El Niño, it became an outbreak and spread as a killing force throughout the Caribbean. The demise of elkhorn coral (*A. palmata*) from the same disease rapidly followed. Then another disease, white pox syndrome, appeared, affecting both species. These diseases devastated populations of these once-dominant Caribbean coral species and drove them onto the endangered species list in 2004. The best studied of these outbreaks was the white pox disease of *Acropora palmata* in the Florida Keys in 1994, initiated by Jim Porter at the University of Georgia.

Jim and his team started photographing thousands of individual *Acropora* corals in 1994 and showed that over the next decade, the average live coral surface area declined from 20% to 4%, and the number of living elkhorn colonies declined from ninety at one site in 1994 to one or two in 2014 (Sutherland et al. 2016). Jim thought white pox disease was part of the decline, and he suspected poor sewage treatment could be a source of infection and warm temperatures fanned the flames of the disease. Jim and his colleague Katie Sutherland worked for over a decade to isolate one causative infectious agent of white pox and track its source. They found that one of the deadly pathogens was a species of fecal coliform bacteria called *Serratia marcescens*. The pathogenic strains affecting the coral exactly matched strains taken from a nearby septic system. They concluded that human sewage was indeed the source of this pathogen.

More work by other researchers has identified additional agents, including protozoan ciliates and possibly even viruses, that can also cause outbreaks of pox-like lesions on coral. Because the role of each infectious agent in producing the signs of disease can shift over time, determining causality is what scientists call chasing a moving target. In some cases there is a primary cause of infection, like the *Serratia* bacteria. In other cases, signs of infection similar to pox spots can be caused by other microorganisms, or even a changing consortium of infections. Added complexity comes from the fact that the infections are made worse by warming events, when the corals are stressed and the microorganism grows fastest. Sutherland and Porter's careful work in identifying the causative bacterium and one of its sources in human sewage is an example of the kind of science that is key to developing solutions for preventing and controlling future disease outbreaks.

The disease outbreaks in sea fans, elkhorn corals, and staghorn corals that occurred in the 1980s and 1990s, along with rapidly declining live coral cover, lit the fire under scientists and policymakers. After the disastrous effects of the 1997–98 El Niño warming event on coral health and the dawning realization that many national and local economies were dependent on healthy reefs because of their role in providing fisheries, tourism revenue, and shoreline protection, I was asked by the World Bank to develop an international working group of scientists to study coral health. This was part of a large international project led by Maria Hatziolis of the World Bank and Nancy Knowlton, then of Scripps Institution of Oceanography, involving five other working groups focused on coral reef sustainability and fisheries that established four nodes of scientific capacity in Australia, Africa, Mexico, and the Philippines. My own research team of seven scientists and their students—Garriet Smith (University of South Carolina), Bette Willis (James Cook University, Australia), Ernesto Weil (University of Puerto Rico), Laurie Raymundo (University of Guam), Farooq Azam (Scripps Institution of Oceanography), Eric Jordan (National Autonomous University of Mexico), and I—worked through three global coral bleaching events, from 1997 to 2015, racing to determine the causes of the death of vast swatches of coral reef as sea surface temperatures rose in the Florida Keys, the Yucatan Peninsula, Zanzibar Island in East Africa, and Wakatobi in Indonesia (see map 1). My husband jokingly referred to us as a band of eco-warriors, based on our focused hard work. But we were all in the grip of a sense of profound loss as we watched coral reefs decline. I still cannot find words to describe how I feel about the tragic loss of these reefs, the most biologically diverse ecosystem on our planet.

I remember the day in Akumal when the full impact of the unraveling coral reefs hit me. It was 2006, the too-warm summer following the devastating category-five hurricane Wilma in the fall of 2005. The year 2005 had been the warmest on record. We had worked on the beautiful reefs of the Yucatan since 1995, when the fore-reefs were still dominated by the giant reef-building elkhorn coral, *Acropora palmata*. The melting away of the Yucatan's living coral reefs from infectious disease had begun earlier, in 1984, with the demise at many locations of the primary Caribbean corals, elkhorn and staghorn. Although the reefs at our site in the Yucatan had stayed healthy longer than others, by 2004 both *Acropora* species were rare, and the dominant reef-building corals that remained were three species of *Orbicella*, forming towering canyons of five-hundred-year-old colonies of green, brown, tan, and red.

I was checking our long-term sites that day, swimming along transects we had monitored for a decade. The waters were unseasonably warm. As I swam in the environment that corals love—sunny, shallow reef waters about twenty feet deep—I saw a huge colony of *Orbicella* that was riddled with giant lesions and had only a small amount of green living tissue left. I stopped to think

Map 1. World coral reef distribution.

about its life. It was about thirty feet tall and twenty feet wide on one side and thirty feet on the other side. It was an ancient colony, perhaps eight hundred years old. This being had been here for a long time, its individual polyps dying after relatively short life spans but replaced by new ones with the identical genome. It had likely begun its existence in the AD 1200s, the post-classic period of the Mayan civilization in Mexico. Mayans would have been on these very shores when this coral was born, since the ruins of coastal Tulum are a mere forty kilometers away. There is something wrong about living through so much of human history only to be brought down now in our time, in this year. As I continued my survey, I saw that all the big old *Orbicella* corals on our Akumal reefs looked the same way—mostly dead, with only a small amount of living tissue. I thought about how magical this place once was, an underwater cathedral created by living organisms and home to so many others. I thought about the exciting work we had done here and all the students we had taught and inspired to study tropical marine science. I felt so discouraged that we had failed to somehow protect this reef. It was a feeling of such helplessness and frustration; there was nothing we could have done in the face of a warming ocean.

The corals I saw dying after the bleaching event of 2005 were decimated by a combination of outright heat stress and lethal opportunistic disease. One of the killers was yellow band disease, which specifically attacks two of the three species of *Orbicella* Caribbean-wide and was the cause of the lesions I saw in Akumal. It is destructive, but also an unusual coral disease because rather than attacking the coral itself, it is thought to specifically target the symbiotic algae living inside the coral. It sickens and kills these normally green algae and causes a characteristic yellow area where the algae are either discolored or expelled from the coral tissue. Under a microscope, the algal cells are yellowish and shaped like disfigured, deflated tennis balls instead of bright brown or green spheres. Eventually the coral itself also dies as the multiple lesions on a colony spread to overtake all the living tissue. Yellow band disease, like many infectious diseases, is affected by temperature. Ernesto Weil, professor at the University of Puerto Rico in Mayaguez and a member of our working group, monitored the impacts of yellow band disease Caribbean-wide and particularly on his home reefs of Puerto Rico. His research showed how the lesions created by pathogenic bacteria increased dramatically in the summer and barely at all in the winters between 1999 and 2003. But after 2003, temperatures in the winter began to warm enough that the lesions continued to grow during the winter as well. By the end of the decade, the lesions were growing much faster in the summer and the winter lesions were growing as fast as they once had in summer. There was no seasonal relief for the corals (figure 4).

What Ernesto documented so well in Puerto Rico was also playing out on the Yucatan and all over the Caribbean, knocking back the then-dominant Caribbean coral. As if it wasn't bad enough in 2006, the outbreaks of disease following the 2005 bleaching event were repeated in the 2010 warming event and again in 2017 on reefs around the world. A National Public Radio headline from May 15, 2018, reads "Battered by bleaching, Florida's coral reefs now face mysterious disease." In that article, Bill Precht says it's proved especially deadly for species of brain and star coral, which form the foundation for many reefs. According to Precht, in some areas almost all of those corals are now dead. "This is essentially equivalent to a local extinction, an ecological extirpation of these species locally," he says. "And when you

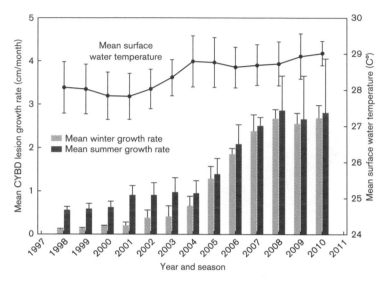

Figure 4. Yellow band disease lesion growth rate in relation to temperature (adapted from Burge et al. 2014).

go out and swim on the reefs of Miami-Dade County today, it would be a very rare chance encounter that you'd see some of these three or four species."

. . .

Managing the World Bank project led me to the coral reefs of Africa, Australia, Micronesia, and Southeast Asia to survey coral health and work with local scientists, providing expertise in developing centers of excellence (see map 2). The work I did in Indonesia feels the most meaningful, because I was able to contribute to surveying the health of those reefs and help train Indonesian scientists. I had been first enticed to Indonesia in 2011 by the mind-bending beauty of the coral reef biodiversity and the chance to work with Indonesian reef scientist Jamaluddin Jompa from Hasanuddin University in Sulawesi, then the director of a large coral

Map 2. Southeast Asia and Indonesia.

conservation program managed by the World Bank and funded by the Global Environmental Fund. Indonesia is in the heart of the so-called coral triangle, home to the richest marine biodiversity on our planet. There are over 580 species of hard coral alone in Indonesia, as compared to 67 species in the Caribbean. There are approximately 3,000 species of fish, as compared to 1,200 in the Caribbean. And these are just the ones scientists have catalogued. Even today, we are still finding new species.

My dives on the unspoiled coral reefs of Indonesia were the most spectacular of my career. My first dives were in 2011 in the islands of Wakatobi National Park, just past Wangiwangi and around the corner from Bau-Bau in southeastern Sulawesi. I was accompanied by my colleague Bette Willis and my high-school-aged daughter. Morgan, already an intrepid traveler, good photographer, and strong scuba diver, was along to provide an extra pair of hands and help as land and underwater photographer. We dropped underwater to a steep drop-off reef, a wall of coral stretching from fifteen feet below the surface to more than two hundred feet. The current was rapid, so this was going to be a drift dive, and by the end we managed to cover nearly a kilometer underwater.

Let's first talk about the fish, because this is how a reef should be, especially one near the center of coral and fish biodiversity in Indonesia. There were so many bright fish at all levels of the food chain. From the green-blue parrot fish and yellow tangs grazing on algae, to the coral-eating butterfly fish, to the spectacularly bright blue and red plankton-eating *Chromis* hiding nervously in the branching corals, to the lurking moray eels, the scene was a multihued circus. And below us in deeper water lurked a cadre of much larger predatory fish that I couldn't even identify. All across the reef flat were giant anemones with their resident clown fish. The makers of *Finding Nemo* had it right when they picked the clown fish as their star—nothing is as enchanting as a family group of brightly colored, alert clown fish nestled restlessly in the tentacles of a giant anemone. The image says so clearly, "This reef is my home."

We were rushed along by the current at a pretty good clip—fast enough that stopping wasn't possible without grabbing for a rock outcrop. The cliff face was covered in rich reef-building corals that are solar-powered by their symbiotic algae, but also a myriad of brightly colored marine invertebrates like sponges, sea squirts, and soft corals that were dependent on the current to bring them food. Lots of overhangs and holes created lobster condominiums, revealed by long antennae poking from the reef matrix. Vodka-clear water, cleansing water currents, deep drop-off reefs, an amazing diversity of brilliantly colored and bizarrely-shaped corals, sponges populated with bright, fast-jetting fish, lurking sea snakes, and brilliant sea slugs all combined to create an astonishing magical environment on most of our dives.

During these dives, our team accomplished the first set of coral health surveys on both sides of South Sulawesi, at Wakatobi National Park on the eastern side and the Spermonde Islands on the western side. Our initial team was composed of Bette Willis (James Cook University), Laurie Raymundo (University of Guam), Erin Mueller (Mote Marine Lab), and Courtney Couch (NOAA, Hawaii). They are all now distinguished coral biologists, but Bette in particular is one of the world's top coral biologists, specializing in coral systematics and reproductive biology and more recently, coral health. It would have been impossible to survey the health of Indonesian coral reefs, composed of such a high diversity of coral species, without her expertise.

During our time in Indonesia, we organized a workshop to train Indonesian scientists on the basics of coral health and survey methods. We stayed at the Hasanuddin University Marine Laboratory on Barrang Lompo Island in the Spermonde Archipelago. This was not the easiest place to run a meeting. Electrical power came from a small generator that was turned off at night and started up again at about six in the morning. There were no flushing toilets—near our room there was a bathroom that had one conventional western-style toilet and one Indonesian style, a hole in the ground with footrests on either side. There were no showers, so we scooped cold water out of a large holding tank and splashed it over our heads. In addition to the challenge we faced with hygiene, our schedule was influenced by the importance of respecting the local culture. Spiritual practices are deeply embedded in the daily routine in a devoutly Muslim culture like that of rural Sulawesi, with prayer services throughout the day. The day started with prayers broadcast on a loudspeaker to the entire island at about 4:30 am; then came pre-breakfast services in the mosque and prayers at noon and in the afternoon. Our Indonesian colleagues attended all these devotional activities. Despite technical difficulties with the generator and frequent breaks for prayer, we were able to successfully equip about twenty Indonesian scientists and students with basic information about the methods and tools to use in the study of coral disease outbreaks, and to convey the background knowledge needed to understand the rationale behind the protocols.

In the afternoon, we took the entire group onto the coral reef to learn coral species and practice field health surveys. One moment is forever branded in my mind. One student said she had never snorkeled below the surface. She was afraid, but once she saw us diving down she was determined to observe more closely the black band disease infection on the coral colonies. My colleague from the Nature Conservancy took her by the hand and said, "Let's do it together." The two of them dove down to the bottom holding hands, with Ocha in a hijab and shawl. To this day, this image stands out for me as a powerful symbol of our quest to train all scientists and especially to empower women wherever we work. Another of our female students said she had not used scuba gear before and, in a gesture of solidarity, one of our male Indonesian students offered to teach her and spent the next two days instructing her. I appreciated that nearly half the participants in our workshop were women, and I am grateful to my colleague Jamal for his role in supporting women students in his group and encouraging us as women scientists to interact with and train his students. He has also made a point of emphasizing the advancement of women administrators at Hasanuddin University and, on every trip, would involve me in meeting female leaders at the university. I respect and appreciate his mission to promote women in science considering the gender inequality found around the world. We came to Indonesia to contribute to conserving nature and additionally learned our value as female role models.

On the last day at breakfast, one of our students was too sick to eat. Soon after, as we left for our two-hour boat ride, I started to feel nauseous and exhausted. Around me, others complained of lethargy. By the time our boat docked in Makassar, we had to run for the public bathrooms at the dock. The bathroom was dark, with a dirt floor and the grimmest-ever lack of hygiene. During the next hour, I got progressively worse. By the time we finally reached the hotel, I was so weak I had to be helped to my room. That night I was so sick that I was literally crawling to the bathroom. I felt a little better by the next morning, but was still weak and sick and troubled to learn that Bette and Erin and Laurie had also fallen ill during the night. After another day of sick travel, we reached a clinic in Bali, where we all tested positive for amoebic dysentery and were put on the dread anti-protozoan medication Flagyl. I've taken Flagyl a few times and hate the stuff. It creates a nasty metallic taste, contributes to nausea, and means no wine or beer for two weeks. But it killed our amoebas. As the rest of us began to recover over the next week, Bette Willis got weaker and weaker. Further tests revealed she was positive not only for amoebic dysentery but also typhoid. Her husband flew over from Australia and took her back home, but it was months before she recovered, and over a month before the rest of us felt one-hundred-percent again. Despite the scrupulous measures we took to protect our health, this wasn't the first bout of illness we would encounter in Sulawesi.

It is common for travelers to Southeast Asia to suffer from intermittent dysentery-like diseases, but it's much less common for people to get amoebic dysentery and typhoid. It was all tragically ironic; in an effort to bolster reef health, we disease ecologists were falling ill. A few people even joked to us about it: "How can you cure corals when you can't even stay healthy yourselves?" At the time, I had no clue how horrendous the hygiene was on that island. Not only were there no real septic systems, but many people on the island regularly defecated on the beaches just outside their houses, beside the marine lab. In 2010, three people from Barrang Lompo died of typhoid. In 2015, there were 29 confirmed cases of typhoid and 192 of other diarrheal diseases requiring hospitalization. In March 2016 alone, 12 people were reported with typhoid. The outbreak among our researchers was a wake-up call and inspired us to design what was to become a landmark study: a project linking coral and human health. We focused on the effects of poor hygiene on the waters and showed how using the ecosystem services of seagrass could make coastal waters cleaner. I will come back to our study on Indonesia hygiene in the last chapter, where I discuss the role of seagrass beds in cleaning up the sewage around that same island. We continue to work with Jamal's group in Indonesia, doing research on the ecosystem services of coral and seagrass and on how cleaning up coastal waters can benefit coral, farmed marine algae, and people. This work is an example of how, in the battle to save coral reefs from climate change impacts, one valuable tool is better local management of reefs. In addition to improving hygiene and treatment of sewage, it is important to control damage from local fishing and better manage reef fisheries to limit over-fishing.

When we were diving in the underwater paradise of Salt River Canyon off St. Croix in 1983, we could never have imagined the global catastrophe that would unfold on coral reefs over the next thirty years, punctuated by the coral bleaching events of 1988, 1997–98, 2005, 2010, and 2016–17, and the massive demise of Great Barrier Reef and Indo-Pacific corals in 2016 and 2017, documented recently by Terry Hughes, director of the Coral Reef Center of Excellence at James Cook University in Townsville, Australia. The feelings of frustration and sadness that arise from our failure to safeguard the health of the world's coral reefs are shared by hundreds of other coral reef biologists the world over. Marine ecologist Peter Sale, author of the 2011 book *Our Dying Planet,* gave the book its grim title because it seemed the most direct statement of what he felt is happening to the earth; for Peter and so many others, nothing is more emblematic of our global ecological crisis than the decline of the magical ecosystems that are our tropical coral reefs.

The ongoing demise of coral reefs is finally receiving attention in the media. Most popular coverage and discussion identifies the cause of coral reef decline as the warming of tropical waters caused by climate change and focuses on bleaching damage as the main cause of death. The role of infectious disease in the decline of coral reefs is rarely if ever mentioned. It is satisfying to see the success of the terrific new documentary film *Chasing Coral,* which tells with spectacular imagery the tale of the catastrophic decline of the world's coral reefs. The film is powerful enough in conveying the sense that we as humans also belong to nature that some of my students cry in despair when they watch it. We can hope this kind of strong filmmaking will continue to inspire action when it comes to climate change. Dire as the story is, I feel the film pulls some punches; while it communicates clearly that warming results in bleaching and the loss of symbiotic algae and the death of vast stretches of coral reef, it leaves out infectious disease. Heat stress and loss of required symbionts are bad for coral and by themselves can cause death, but this combination of stresses also leads to opportunistic infections that continue the killing. Without disease, the bleaching events would be less lethal and fewer corals would die in the years between events. The world's coral reefs, our most diverse and valuable marine ecosystems, are being sickened by a variety of factors all at once. Our only chance to limit their loss lies in understanding how all the threats to coral interact with and affect each other.

CHAPTER THREE

Abalone Outbreaks

A Steady Path to Extinction?

When you believe you have found an important
scientific fact, and are feverishly curious to publish it,
constrain yourself for days, weeks, years sometimes,
fight yourself, try and ruin your own experiments,
and only proclaim your discovery after having
exhausted all contrary hypotheses.

Louis Pasteur

In the mid-1980s, Brian Tissot was a graduate student studying black abalone on Santa Cruz Island, one of the Channel Islands off Santa Barbara, and Año Nuevo Island, off the California coast south of San Francisco. He was a dedicated surfer, diver, and adventurer and looked the part of your basic bearded, tanned, 1980s surfer dude. He is now director of the Humboldt Marine and Coastal Science Institute. The initial goal of Brian's abalone study was to understand how the shape and size of black abalone shells varied among populations along the West Coast. He had noticed that not all abalones had the same shape shells or the same number or shape of respiratory holes, and he wanted to understand whether degree of wave exposure, growth rate, or type of food governed shape and size. So he needed to tally the numbers of different shapes and sizes of shells. Black abalones are the only California species that live in the intertidal zone, so Brian was able to complete his surveys without diving. It's a good thing, because Brian was working on islands surrounded by shark-infested waters. Año Nuevo is part of the "red triangle" off California's coast where most shark attacks occur. When Brian started his intertidal surveys, the abalones at Santa Cruz Island were so dense they piled one on top of the other, so he had plenty of study animals at the beginning. He would stretch his transect tapes across the rocks and count every abalone and its size and shape near the line. Intertidal surveys may sound like a breeze, but these off-shore islands are remote, treacherous and wave-bashed. Many of Brian's transects were located on cliff edges.

The black abalone were so dense on Brian's eight transects on Santa Cruz Island that in the beginning, it took him three days to count them all (see figure 5). He told me he remembers wondering if it was pointless to try and count so many. But then on one trip, when he returned after being delayed by a big winter storm in 1988, he was surprised to see open space and many empty scars in the rock on some of his transects. A lot of abalone had suddenly disappeared, and he did not know why. The counting was suddenly easier and very relevant; Brian thought that maybe only certain sizes or types had died. At first the change was only apparent on some transects, so he thought perhaps the storm and scarce food had weakened those animals. Over the next months, however, the numbers of abalone continued to decline on all his transects, to the point where he could finish all his counts in a single day. He also noticed some in the process of dying: the foot of a dying abalone lost muscle tone and mass,

Figure 5. Abalone at the same site on Santa Cruz Island, photographed in 1986 and again in 1988, at the peak of black abalone mortality. Photo by Brian Tissot.

weakening until the abalone could no longer hold onto the rock and fell off and died. When he wrote his PhD thesis in 1990, after monitoring over 2,715 tagged abalone on Año Nuevo and Santa Cruz, Brian concluded that abalone were dying because the 1986–87 El Niño had caused a huge food shortage and they starved. By 1990, 95% of the black abalone were gone from his transects on Santa Cruz, and biologists had noticed the decline elsewhere. Brian felt depressed about the large-scale die-off and bothered by the realization that something big was going on that he didn't understand.

Abalones are primitive marine snails that have single, flat, ear-shaped shells instead of the tall, coiled ones we normally associate with snails. The adults of some species are larger than dinner plates and can live thirty years or more. The flat shell, with its distinctive set of four to eight respiratory holes on top, covers the abalone's body and its huge muscular foot. Like other snails that eat plants, they are well equipped to deal with the tough cell walls of red and green algae; they have large grinding mouthparts edged with sharp, slender teeth, and hugely long intestines for thorough processing. Baby abalone with their tiny pink or dark shells and questing antennae possess an undeniable charm. No bigger than ladybugs, they trundle around grazing on tiny single-celled algae stuck to the rocks before graduating to eating baby kelp and other ocean plants. When fully grown, they feed on larger kelp and drift algae.

Abalones live along the rocky shores in tropical and temperate waters of the world and all told there are approximately fifty-six species. California once boasted the second-highest species diversity in the world with a full rainbow of seven or eight different species (depending on whether you are a taxonomic lumper and want to combine taxa or a splitter and want to divide taxa into more different kinds). Australia tops California with eleven species. The California abalone species run the gamut from the quite spectacular dinner-plate-sized red abalone, the largest in the world, to the more diminutive pinto abalone. The pinto is the only species with a broad enough range to also frequent the waters around my home in the San Juan Islands. Between the red and pinto in size, in order of historic abundance in California, are the black abalone, white abalone, pink abalone, green abalone, threaded abalone, and flat abalone (see figure 6). The shells of different species differ in color and texture and in number of respiratory holes, but also in the rather fabulous color of the iridescent mother-of-pearl coating their inner shells. The inside of the green abalone shell is a lustrous blue-green mother-of-pearl prized by jewelers. Prized also—for eating—is the abalone's large muscular foot. They were harvested avidly by Native Americans for jewelry and trade and, until relatively recently, West Coast recreational divers. They are also farmed across the Pacific. Abalone fetch a high price ($150 a pound) on Amazon and more in the Asian seafood market.

· · ·

In his PhD thesis, Brain Tissot reported that the decline of black abalone in California began about two to three years after the super El Niño year of 1982–83. That was a hot year, with temperatures soaring around the globe and weather-related disasters recorded in Australia, Indonesia, and Africa. On the US West Coast, the ocean was 3.5°C warmer than the previous average. Further north, in Washington State, I was a graduate student and we were eating tuna, due to an unusual northward pulse of tuna with the warming. As biologists in California became aware of the abalones' lower numbers, they proposed

"Red" Abalone
Haliotis rufescens. Average size: 8"–9"

"Pink" Abalone
Haliotis corrugata. Average size: 5.5"–6.5"

"Black" Abalone
Haliotis cracherodii. Average size: 4.5"–5.5"

"Flat" Abalone
Haliotis walallensis. Average size: 4"–5"

Figure 6. Eight abalone species from California. Photo by Buzz Owen.

"Green" Abalone
Haliotis fulgens. Average size: 6"–7"

"White" Abalone
Haliotis sorenseni. Average size: 6"–7"

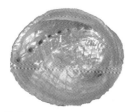

"Threaded" Abalone
Haliotis kamtschatkana assimilis. Average size: 4"–5"

"Pinto" Abalone
Haliotis kamtschatkana kamtschatkana. Average size: 4"–5"

several explanations: the abalone had migrated to cooler waters, they had been over-harvested, or they were dying from heat or starvation stress. But as the years went by and abalone populations continued to diminish and abalone were observed with shrunken, withered feet, it became clear that the problem was not migration or over-harvesting. Brian continued to monitor the abalone for fourteen more years, hoping to document the recovery that never occurred. After he landed his dream job at University of Hawaii at Hilo, he then focused more of his research on science supporting Hawaii's coral reefs and fisheries ecology, but he continued his annual survey and wrote the report that contributed to listing black abalone as endangered.

In 1989, a few years into the decline, Carolyn Friedman, now a professor at the University of Washington, was then a young biologist newly hired as a fisheries pathologist by the California State Department of Fish and Game. She was assigned the task of figuring out why abalone were dying. Although small in stature, Carolyn is a giant among scholars and tenacious as a researcher. She and her coworkers (notably her good friend and abalone biologist Pete Haaker) were confident that some kind of infection was the cause. They coined the name *withering syndrome,* or WS, for the disease, in reference to the withered foot of affected animals. Ron Hedrick's group at the University of California at Davis had recently suggested a parasite as the causative agent, so Carolyn first followed up on this hypothesis.

It was hard to know where to look for causes of the unusual mortality, since there was no baseline of abalone health nor previous images of their tissues. Carolyn started by running the same kinds of autopsies on sick and dying abalone that human pathologists run on sick and dying humans. She noticed the same thing that Brian Tissot did: the feet of dying abalone were losing tone and mass, then the abalone fell off the rocks and died. Carolyn wanted to know what was happening inside the sick abalone. She dissected the animals and checked the health of each internal organ. She made histological (tissue) slides of all the organs to look for tissue abnormalities or infectious agents under a microscope. In slides of the kidneys from pinto abalone, she noticed a protozoan parasite called a renal coccidian. Coccidians are common destructive parasites in mammals; examples on land are *Toxoplasma gondii* in otters and cats and *Sarcocystis* in horses, dogs, and possums. Without a baseline for the health of abalone, there was no way for Carolyn to know if the coccidian was a new health threat or not.

Carolyn looked at more slides of the black abalone and found extensive infections by the coccidian. To test for it as the infective agent, she needed to use abalone that she knew had never been exposed. She used pinto abalone from hundreds of miles north of where the abalone were dying, exposing them to the protozoan—which she had described as a new species—to see if the infections would reproduce the clear signs of disease. In her experiments, the protozoan infected the pinto abalone but didn't kill it or cause the foot to wither, so she concluded it was not the infectious agent and went back to the hunt for the killer.

Carolyn had also been seeing large purple spots, called inclusions, when she looked at slides that had been made from the digestive tracts of sick abalone. From the inclusions she identified microbes that she confirmed as rickettsial bacteria. The inclusions showed up in the other species she was studying, the black and red abalones. What was confounding was that she found purple inclusions in the tissues of apparently healthy abalone in the north. The range of abalone with the purple inclusions in their tissues far exceeded the range of the sick abalone.

Because of this mismatch, the rickettsial bacteria seemed like another dead end.

While Carolyn was examining tissue slides and running her experiments, a cadre of biologists was surveying black abalone up and down the California coast. Jessica Altstatt, Pete Raimondi, Kevin Lafferty, and others produced survey data that showed a striking geographic pattern in the disappearance of the abalone. Jessica's paper showed that the big declines began in the south and spread northward. Pete Raimondi, then a graduate student and now director of the Institute for Marine Sciences at the University of California at Santa Cruz, published a paper in 2002 with a devastating figure of black abalone decline that showed the potential roles of both the 1997–98 El Niño and the earlier smaller one of 1991–93, which hit harder in the south. (See figure 7 and map 3.) At the southernmost survey site, Government Point near Point Conception, nearly all the abalone disappeared during and after the first El Niño, between the spring of 1992 and the fall of 1993. A little further north, at the Boathouse and Stairs sites, abalone numbers held through the early El Niño but began to decline soon after, in 1995, and crashed to virtually zero by the spring of 1998. Even further north, at the Cayucos and Purisima Point sites, the population didn't begin to crash until the fall of 1998, after the large El Niño hit. At Piedras Blancas, north of Morro Bay, black abalone numbers held strong through 2001. By 2001, black abalone had completely disappeared at all sites south of Cayucos Point, in the southern part of their range. This latitudinal pattern through time is a vivid snapshot of a temperature-sensitive epidemic. Jessica Alstatt's group called this local extinction of abalone at many southern sites a geographic extirpation.

As part of her ongoing investigation, in the early 1990s Carolyn Friedman visited Santa Rosa Island with scientists from US Fish and Game and the Channel Islands National Park. Santa Rosa Island was within the southern portion of the black abalone range, where abalone were initially experiencing high mortality. Carolyn was impressed by the developing scale of infection. There were still lots of abalone on the shore, but they were weak, with a shriveled foot, and were easily pulled off the rocks. At this point, the early survey data and her own observations on the pattern of spread had convinced her that the withered foot and mass mortality were the result of an infectious disease, even though her experiments with the coccidian parasite were not showing a link to WS. On Santa Rosa, Carolyn's friend and former college roommate Jenny Dugan (now a professor at UC Santa Barbara), took a photograph over Carolyn's shoulder of two black abalone suffering from WS. In this now-iconic photo, the abalone on the right, in the early stages of the disease, has a large foot that fills the shell; the animal on the left has the withered foot of a sick abalone (see figure 8).

In the late 1990s, although no relationship had yet been established between the geographic extent of the sick abalone and that of the rickettsial infections, Carolyn decided to try transmission experiments with the rickettsia while she continued work on the physiology and immune capabilities of the abalone. She and her students set up black abalone in flowing seawater tanks at Bodega Bay Marine Lab and exposed them to rickettsia released from infected abalone held in header tanks above the experimental animals. Then they waited. After eight months, some of the abalone in the experimental tanks developed a withered foot. Just when it seemed that a promising result was at hand, a boiler malfunction caused the seawater in the tanks to heat up and the experiment was destroyed.

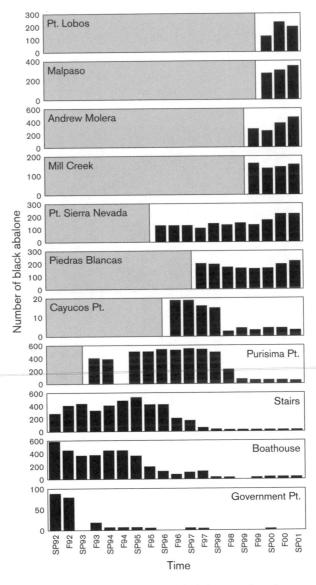

Figure 7. Ecological surveys of black abalone numbers from south, at the bottom, to north along the central California coast, 1992–2001 (adapted from Raimondi 2002).

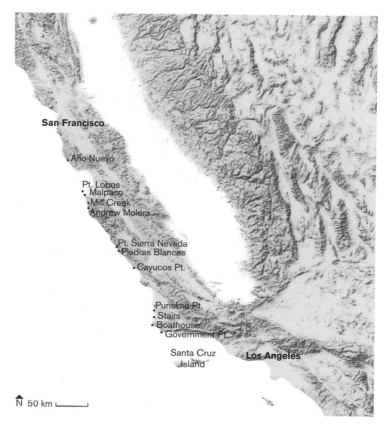

Map 3. Central California study sites for abalone surveys.

Carolyn started a new series of experiments with the rickett-sia, this time with more tanks of infected abalone. After six months, the abalone in her experimental tanks started dying. The abalone infected with the rickettsia finally developed WS. After more than eight years of challenging experiments, each of which lasted one to two years, she had cracked the case: a very slowly incubating rickettsial bacterium was killing the abalone.

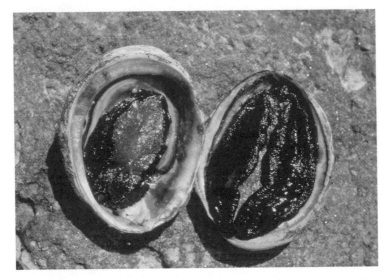

Figure 8. Comparison of the withered foot of a sick abalone on the left with a healthy abalone on the right. Photo by Jenifer Dugan.

Being the thorough investigator she is, Carolyn conducted additional infection trials for several more years to assemble conclusive evidence confirming her hypothesis that the rickettsial bacterium causes WS. Years of additional experiments may sound like a lot of extra effort, but firmly establishing a causative relationship between a microbe and a disease in the ocean is harder than finding a needle in a haystack. This was a particularly hard case because the rickettsia is a very slowly incubating disease agent. It took almost a year before the abalone were conclusively infected. The long incubation period was also why the range of the pathogen initially looked much larger than the range of sick abalone: once infected, it took more than a year— and often years—for an abalone in nature to show signs of the disease, and even longer for it to die.

Carolyn published a paper in 2000 with clear evidence that the infectious agent of WS is a rickettsia-like organism (WS-RLO) that infects abalone gastrointestinal tissue. She named the pathogen *Candidatus* Xenohaliotis californiensis. This sneaky parasite slowly disabled the digestive tissues of the infected abalone, starving it and causing the foot to wither. The long road to identifying the rickettsia would prove to be just the beginning of this story, since there were unexpected twists in the interaction between abalone and the rickettsia still to come.

．　　．　　．

Fast-acting virulent diseases—like our starfish killer—are scary to watch since prime-of-life animals can suddenly be killed in large numbers. But slowly incubating diseases like abalone WS are just as scary, and their causes can be even more difficult to identify. An example of a slow-rolling human disease is Creutzfeldt-Jakob prion disease, a rare and fatal brain disorder linked with bovine spongiform encephalopathy, or mad cow disease. The most likely cause of Creutzfeldt-Jakob disease is infection by a prion, a misfolded protein particle; people who develop the disease are exposed by eating food contaminated by infected cattle brain or other central nervous system tissue. The disease has been difficult to pin down since it may take ten years for symptoms to appear.

The causal agent of abalone WS has been identified, but it is unusual among bacteria and we don't know very much about it. On land, rickettsial bacteria are never free living and can survive only inside the cells of a host. Scientist call them obligate intracellular parasites. They are always transmitted by a bite from an insect vector like a tick or a mite. Two diseases of terrestrial animals caused by rickettsial bacterial are Rocky Mountain spotted fever and typhus. It is tricky to confirm a rickettsia as the cause of a disease, since they hide inside cells and thus cannot be detected by a simple blood test.

Given that rickettsia are intracellular and on land require a vector—the insect—to move among hosts, how marine rickettsia spread among abalone was a puzzle. Carolyn explained to me what she has figured out: since seawater has essentially the same salinity as abalone blood, the bacterium can survive briefly outside a host and can thus be transmitted in seawater. She and her student Lisa Crosson tested infectivity and found that the rickettsia remains viable in seawater for at least twenty-four hours. It is extraordinary that this rickettsial bacterium deviates from the normal transmission biology of requiring a vector on land. Direct transmission in seawater is much faster than via a vector. This is a striking example of why we have more problems with disease transmission in the sea than on land. Seawater facilitates survival and the spread of pathogens, while air does not. Imagine if the diseases that require vectors like mosquitoes in order to proliferate could be transmitted directly, spreading freely in the air? If the protozoan that causes malaria or the virus that causes dengue suddenly did not need to be transmitted by mosquitoes to move from person to person? It would be impossible to control these diseases. We would be in big trouble.

The abalone are in big trouble too—but maybe the situation is not hopeless. In 2002, something unexpected began happening with the abalone in the Channel Islands. Glenn VanBlaricom, a researcher at the University of Washington, noticed that some black abalone were surviving at one site on the remote San Nicolas Island, about sixty miles off the California coast. It seemed that these abalone might be resistant to WS. Carolyn had wanted to test if some black abalone were resistant, so they worked together to collect young abalone from the island and from the mainland at a site without the disease (Carmel, California). Carolyn and Lisa Crosson then ran another long-term experiment. They found that when exposed to the rickettsia, the San Nicolas abalone survived longer and in greater numbers than abalone that had never been exposed to the disease. This was an exciting find; some abalone now appeared resistant to the rickettsia. Years of experiments later, in 2014, Lisa and Carolyn concluded that the San Nicolas black abalone were in fact becoming more resistant than abalone from the central California mainland. It appeared that some San Nicolas individuals had a genetic variation that gave them increased resistance, and through natural selection these survivors were producing similarly resistant offspring.

But there was an added twist. When Lisa and Carolyn double-checked microscope slides from their newer experiments to be sure the abalone were truly infected, they were puzzled at what they saw. The abalone still had large purple inclusions of rickettsia in their digestive systems, but there was something extra in the pictures of the abalones' intestines: big dark blue inclusions. Carolyn had seen something like this before and thought a new parasite had entered the picture. When she used an electron microscope to look at these inclusions, she saw that the rickettsia had themselves been infected—by a virus. Viruses that infect bacteria are called phages. The abalone were developing resistance through natural selection *and* a phage was disrupting the pathogenicity of the rickettsia, basically augmenting the abalone's disease resistance. This was welcome news, but it introduced a new puzzle: where did the phage come from?

Carolyn speculates that the rickettsia were introduced to California without the phage in the early 1980s via infected abalone from somewhere else. Once the pathogen appeared on the California coast it survived well enough to begin its spread and was probably helped along by the 1982–83 El Niño. Then, years later, the phage appeared. But it is also possible that the phage was introduced with the rickettsia and then evolved so it could infect the rickettsia and begin to slow the progression of this disease. Either way, the presence of the phage may explain why abalone farms are not seeing as many animals dying during warm-water periods as they had before the phage emerged. The phage now appears to have spread throughout much of California where abalone are found.

Even though the development of resistance among surviving abalone and the spread of the rickettsia-infecting phage may give abalone populations a chance to recover over time, the rickettsial bacteria have a powerful ally that may give them an edge: climate change and the slow warming of the oceans. Several pieces of circumstantial evidence point to the possibility that a warming ocean environment fueled the outbreak and continues to benefit the rickettsia. In 1988, the first place that WS came ashore after the Channel Islands was at Diablo Canyon, where the waters were anomalously warm due to effluent from the Diablo Canyon nuclear plant. In an email, Brian Tissot said, "I've always envisioned the WS progression as an introduction of an exotic pathogen in SoCal, the spread by water currents, then slow pathogenicity triggered by higher temperatures. I did thermal studies on black abalone at Diablo Canyon in the early 1980s and half of them died above 26C degrees (i.e., they had a lethal temperature that kills 50% or $LT_{50} > 26C$), which made sense as they are big, black animals exposed in the intertidal zone at low tide. Similar studies repeated by Steinbeck and myself in 1988–89 showed a LT_{50} of 18–20C." These results, combined with the fact that the outbreak coincided with El Niño–associated warm-water conditions, supported the hypothesis that warm temperature was adding fuel to the epidemic.

In 1998, a postdoc in Carolyn's lab, Jim Moore, ran infection experiments at different temperatures. He showed that elevated temperatures stimulated development of WS. Kevin Lafferty and Tal Ben-Horin ran additional experiments showing that not only warm temperatures, but also temperature fluctuations increased infection risk. This could explain why intertidal black abalone, the species living in the marine environment with the warmest temperatures and rapid temperature changes, were so devastated by the outbreak.

Carolyn and Lisa looked more closely into the combined effect of temperature and infection on the different species of abalone. They selected three species with different natural temperature tolerances: pinto or northern abalone (*Haliotis kamtschatkana*), with a cool temperature range; red abalone (*H. rufescens*), with an intermediate temperature range; and pink abalone (*H. corrugata*), with a warm water range. This was the first study to test northern abalone, and it showed they were the most susceptible and least resistant when exposed to temperatures that allow infection. Interestingly, by combining data from other studies, Carolyn and Lisa were able to show an evolutionary basis for susceptibility to the rickettsia bacterium. White abalone are the most susceptible, and the degree of susceptibility is highest for those species most closely related to the white abalone, such as the pinto (see figure 9).

Warmer temperatures may not be the only factor that helps the WS rickettsia infect more abalone. In 2012, black abalone in the seawater system at the University of California at Santa Barbara were infected with the WS pathogen. The UCSB seawater

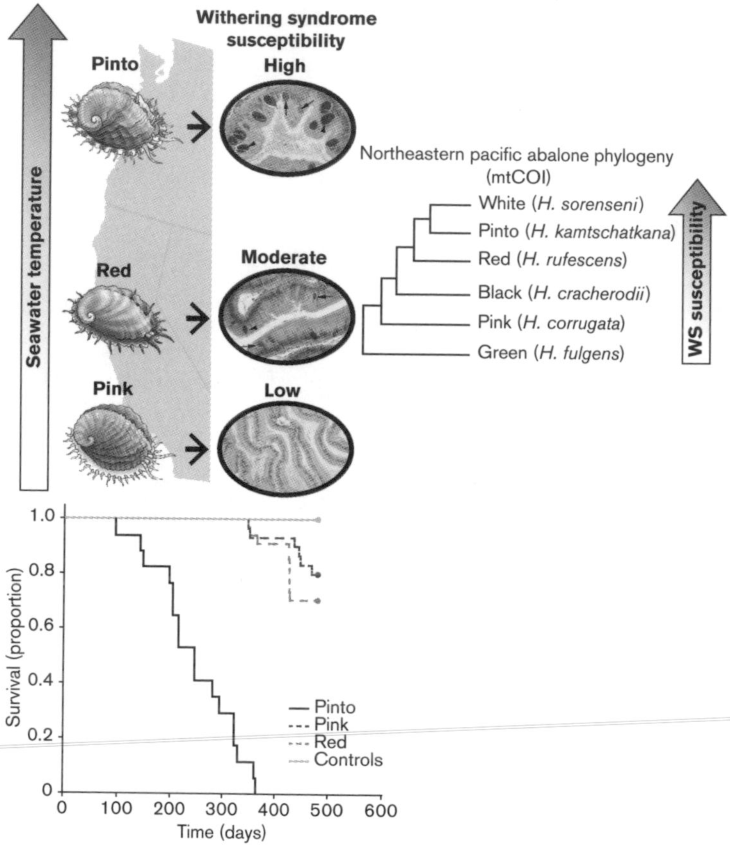

Figure 9. Comparison of susceptibility and resistance to withering foot syndrome disease of three species of abalone. From Crosson and Friedman 2018.

intake is in a place with no known wild black abalone for many miles, and very few red abalone. If the rickettsia is an intracellular bacterium that cannot live for long periods outside its host, how were the abalone in the UCSB tanks getting infected? Was there some other species of mollusc acting as an alternative host, or was there an environmental reservoir? Kevin Lafferty knew that there was an abalone farm not far away. Water pumped into a farm can contain a pathogen like the WS rickettsia, and waters from a farm or aquarium with infected abalone can, in turn, be released back into the ocean along with the pathogen. Kevin decided to investigate the discharge from the abalone farm. Sure enough, there was a high concentration of rickettsial DNA in the discharge, which was rapidly diluted offshore. Curiously, Kevin and his colleagues could also detect rickettsial DNA in samples from more than twenty kilometers away. They published these observations as evidence that the source of the pathogen in the UCSB seawater system was the abalone farm up the coast.

Creating a heated but friendly debate, Carolyn, Lisa, and Ava Fuller contested Kevin's conclusion. They didn't refute that effluent from a facility holding infected abalone contained the rickettsia, but they felt that because populations of wild abalone remained in the Southern California Bight, and water from the islands where many abalone were still found circulated in the area, it was impossible to say for sure that the source of the infection in the UCSB system was the abalone farm. Whatever the source of the UCSB infection, the existence of abalone farms on the California coast raises the concern that commercial farms could be a source of new infections and that wild abalone, in turn, could be sources of new infections in the farms. This problem of pathogen spillover from farmed to wild species and spillback from wild to farmed animals is not just a problem for abalone, nor is it just an issue in California. The next chapter will discuss the spillover of pathogens as it pertains to salmon.

Despite all we've learned since the start of the outbreak more than three decades ago, three abalone species now face a precarious existence. Once a thriving wild and farmed fishery, black and white abalone are endangered in California. The white abalone was listed as endangered under the Endangered Species Act in 2001. In the most recent surveys by the National Marine Fisheries Service, white abalone populations had declined from historic levels of millions to less than 2,500. Scientists classify this as an effective population size of zero because the surviving white abalone are too few to breed and have failed to recruit enough new individuals for the population to recover. In other words, white abalone are on the verge of extinction. Due to precipitous declines, black abalone were listed as endangered in 2009. Black abalone are considered locally extinct at most mainland sites south of Point Conception, California. The continued presence in California waters of the rickettsia that causes WS impedes recovery of these abalone species. The National Marine Fisheries Service is preparing a five-year review of black abalone and white abalone to determine if anything can be done to save them from extinction. The rickettsia-infecting phage and the existence of resistant individuals provide some hope for the black abalone.

Pinto abalone are the only abalone in our inland waters and the San Juan Islands are the region of historically healthy populations. When I was a graduate student in the 1980s, I would always see pinto abalone on dives and in the low intertidal. By 1994, they had declined to the point that the sport fishery closed, and they have not recovered. Their populations are so low in the waters of Washington State—mostly from over-harvesting—that they are listed as a species of concern and have been declared functionally extinct in this locality. Proposals to list the pinto as endangered have been rejected because they thrive in Canada and Alaska and are already protected in Canada. Most of the range for pinto abalone lies north of where temperatures currently cause the rickettsia to be destructive (above 17°C), so it is unknown to date if disease has had an episodic role in their decline. But Lisa and Carolyn's study shows the potential for a susceptible, at-risk species sitting right on the edge of a temperature-sensitive pathogen's expanding range to be trouble waiting to happen in a warming ocean.

.   .   .

The outbreak of WS has been devastating for California abalone, but they represent only a tiny portion of marine biodiversity and an even tinier share of the food we glean from the oceans. Why should we worry about this particular disease outbreak? In a general way, the abalone story is important because it shows so clearly the role that a disease outbreak can have in nudging ocean species to the edge of extinction. If a disease can nearly wipe out abalone, what about other molluscs we eat, like scallops, oysters, and clams? Further, there's nothing unique about California that suggests than any other locality is less susceptible to a similar outbreak.

In a demonstration of the latter point, abalone on the other side of the Pacific Ocean have also been threatened by disease. In the summer of 2006, thousands of abalone died suddenly in a southern Australia abalone farm. The scourge of rapidly dying abalone spread to other farms. The following year, millions of wild abalone in nearby waters died, as recorded by the World Organization for Animal Health. The suspected spread of a pathogen from farmed to wild abalone initiated claims of damage to the wild and triggered lawsuits. The rapid onset of the disease was very different from the slow development of fatal signs typical of the WS experienced in California. The foot of an abalone affected in the Australia outbreak would almost immediately curl at the edges, and it would fall off the rocks and die within two days of showing initial signs of the disease.

After intensive analysis by several research groups led by K. Ellard and C. Hooper, in 2007 a herpes virus was suggested as the causative agent. Several years later, using transmission electron microscopy, a group led by K. Savin showed herpes virus–like particles in the nerves of affected abalone. This was a clear demonstration of the herpes virus as cause. The disease was named abalone viral ganglioneuritis because the virus attacks the nervous system. Most people know about *Herpes simplex* of humans because it can cause cold sores and a sexually transmitted disease. What is less well known is that there are thousands of different virus species belonging to the herpes group, and they are usually host specific, so the herpes virus of abalone cannot be transmitted to humans. Recent work from Australia shows that five different herpes virus variants cause mortality in three different species of abalone (blacklip, greenlip, and brownlip), and there is growing concern that it will spread from the waters in South Australia across the Bass Strait to the valuable farmed abalone in Tasmania.

Another reason to be concerned about outbreaks that may eradicate ocean species is that many species have value beyond that which comes from our ability to eat them. At the very least, every species has an ecological role in the complex food webs of the ocean. The loss of a single species can impact many others, as we saw with starfish. In addition, sometimes animals in the ocean can provide very direct nonfood value to humans. Abalone are a case in point. It turns out that the abalone immune system holds secrets that may prove valuable for human medicine.

In Australia, even before the herpes virus outbreak among abalone there, research was under way on an unusual and new-to-science component of the abalone immune system that is carried in abalone blood. Abalone have blue blood; the color is caused by a giant copper-containing chemical, hemocyanin, which has a super capacity to transport oxygen. It functions like the oxygen-carrying hemoglobin in our own blood. The healing properties of abalone blood were discovered by accident, says Adrian Cutherbertson, CEO of Marine Biotechnologies Australia. In anti-cancer clinical trials, abalone blood serum unexpectedly caused reductions in herpes virus cold sores in volunteers. He also said that an employee responsible for loading and unloading abalone shell containers reported that the viral warts that had plagued his hands for years disappeared after long-term handling of abalone. Further work by V. Dang confirmed the surprising result that the blood pigment hemocyanin reduces activity of the *Herpes simplex* virus of humans. It is currently being developed as a drug that reduces cold sores and warts caused by this virus.

Antiviral compounds have also been detected in other cultured molluscs, including oysters and mussels. The enormous diversity and disease-fighting adaptations of molluscs suggest that many of them are potential sources of novel compounds for future drugs. If species like the white abalone and black abalone are driven to extinction by disease, then we will lose important potential sources of drugs that could benefit humans in our own fight against microorganisms. The opportunity to discover valuable and novel adaptations and mechanisms of defense against pathogenic agents within the biodiverse fauna of the oceans is a theme I will elaborate on in the concluding chapter.

The unusual infectious disease of California abalone has changed the course of an entire species complex in the short span of three decades. It warns us that dangerous pathogens can appear suddenly, spread widely, threaten endangerment, and cause mortality so slowly that unmasking them as causative agents can take many years of hard work. The outbreak also provides another example of how climate change can fuel an epidemic in the ocean and how aquaculture can increase the disease risk for animals both on land and in the sea.

CHAPTER FOUR

Salmon Outbreaks

Food from the Ocean Imperiled

If the salmon and steelhead are running, then as far as
I am concerned, God knows that all is well in His
world.... The health of the environment is good if the
salmon and steelhead are around.

   Former Oregon Governor Tom McCall

I am going out salmon fishing for the first time, with my husband,
Chuck, two old friends, and a fishing guide, Kevin. The morning
is still as we settle into Kevin's decked-out thirty-foot boat, cof-
fees in hand. As other boats gear up, we chug out of the small,
quiet port of Friday Harbor, motor past the Friday Harbor marine
labs, and cross San Juan Channel toward neighboring Lopez
Island. The sun, as usual, is obscured by clouds; even though it is
late July, we need our coats and warm hats. Kevin gets to work
setting up four lines and clips them into downriggers. Two are
dropped to 120 feet and two to 55 feet. We are after Chinook, also
called king salmon and certainly deserving of the name. Southern
resident killer whales (orcas) and humans alike prize Chinook for
their size (easily over 40 and up to 130 pounds), high omega-3 fatty

acid content, great taste, and firm pink flesh. We join a small cadre of boats trolling along the shore at high slack tide.

Kevin explains what will happen if we hook a fish. We decide Chuck should have the first crack at it since our friends have fished before and I am too nervous to be first. In short order, we notice a rod bending and shout to Chuck to land his fish. It's fighting! It's not easy to pull in, but Chuck knows the drill. Then it thrashes near the surface and Kevin gets the net ready, shouting to Chuck to back up. We all look sharp as it comes beside the boat. Then we see it's not a Chinook, but a small shark called a dogfish, common in our waters. They can be up to four feet long and live a hundred years, but they are no Chinook. We are using barbless hooks, so it is easy for Kevin to pull it free and we watched it swim away. We laugh at our big catch, still happy that we hooked something. The lines go back in the water, zipping down to the depths. After some discussion about whether Chuck's turn is over, we decide he can have another chance. The lazy morning continues; fishers in nearby boats have caught Chinook, so we are hopeful. Again, the pole on our boat dips and we spring to action. Surely it's a big fish on the line. Chuck reels it in. What luck—another dogfish. At least it's good to know there are plenty still in our waters. We troll the inland water for another hour with no luck and then decide to head out to the main fishing grounds on the outer side of our island. We motor through the often turbulent but now quiet waters of Cattle Pass, to what might be the most beautiful place in the world.

On the western side of Cattle Pass we are beside the rough, rocky shores of San Juan Island, close to the crossing of Haro and Juan de Fuca Straits. This spot is framed by the snow-capped Olympic mountains across Juan de Fuca Strait and Vancouver Island to the west and north across Haro Strait. To the

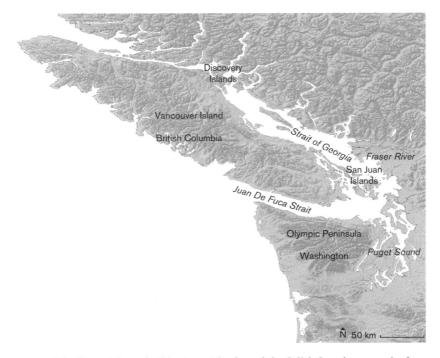

Map 4. The Fraser River, the Discovery Islands, and the Salish Sea, the network of coastal waterways that includes the Strait of Georgia, Puget Sound, and the Strait of Juan de Fuca.

south we can sometimes see the distant peak of Mount Rainier (see map 4). It feels wild, beautiful, and limitless. As our lines go in the water, it feels like something exciting will happen. What will we catch? This is also the route of the orcas, so one could surface alongside. We continue to troll and start to get bored, since nothing much is happening. We don't immediately hook anything and we don't see an orca.

After four hours of trolling back and forth, we are ready to give up. Now Kevin mentions it is a bit early for the summer run of Chinooks, and given the absence of orcas, it is no surprise we are not catching any. If there were lots of Chinook around, we would see orcas busy feasting, since Chinook are their preferred food. We get ready to quit for the day, but suddenly the deep rod dips and the line zings again. Something a hundred feet down has grabbed our lure.

It's my turn to pull in whatever is on the end of the line. I start to reel, surprised at how difficult it is. "No hurry," Kevin says. "Relax and let him fight; just keep tension." David is laughing, "Don't horse it!" I reel slowly and feel the play on my line as the fish rushes deeper, and I reel faster as it rushes up. I do not want to lose my catch, even if it is another dogfish. Then it breaks the surface and we all yell. A beautiful Chinook salmon! It zooms toward the boat and Kevin jumps to keep it free of the motors. The pressure is on to get it in the boat. I back it to boat-side and Kevin nabs it in the net: a gleaming, exquisitely beautiful fifteen-pound fish, silver with bright blue-green speckles. This is not how they look in the fish markets. My excitement edges to regret. I have just hooked a wild Chinook salmon, listed as an endangered species, an extraordinary marvel of nature and prey for our starving orcas, which are also listed as endangered. Expecting to be met with resistance, I say quietly, "I might need to throw it back." Kevin laughs. "You *have* to throw it back, it's a native." My fish has an intact, unclipped adipose fin, which means it was born and bred in nature and not a hatchery. I am happy to have caught it and even happier to release it back to its waters. It is a relatively small Chinook anyway, and Kevin thinks it has a good chance to survive. All in all, a satisfying outcome.

We'd known it was a low year for Chinook before we planned the trip, but we were hopeful nonetheless. Marine stocks are notoriously variable; we'd seen big returns in other years. We were fishing not more than sixty miles from the mouth of the

Fraser River, North America's longest dam-free river and the nursery for the largest run of wild salmon in the world. The Fraser River has a sockeye salmon run with a historical average of eight million fish worth over $1 billion. Here in the San Juan Islands, we are occasionally inundated by big flushes of fresh, cold water from the Fraser that can carry along a big bounty of salmon. The west shore of San Juan Island where we live is basically a cliff against which all that water and fish wash up. In 2014, we kayaked off our beach into a massive run of Fraser River chum salmon. The water literally boiled with jumping fish, so many we were sure one would land in our laps.

.   .   .

A salmon begins its life in a freshwater stream, grows to maturity in the ocean as a fierce, wily predator, and returns to its natal stream to spawn and die. Let's follow the anadromous life of one sockeye salmon from the Fraser River, which accounts for more than half of the sockeye in British Columbia. Our fish starts as an egg buried in sediments of the river near Stuart Lake, eight hundred kilometers upstream from the ocean. When she hatches under the gravel, she is an inch-long alevin and still has an attached, nutrient-filled yolk sac. She lives tucked in the gravel, sustained by her yolk, for a few months. During this time, she imprints on the locale, forming chemical memories of the site. In late spring, following an ancient pattern stored in her genetic code, she burrows out from the gravel and migrates to a nearby lake, where she lives for one to two years. Then she makes her way downstream, entering the big ocean as a young smolt, sleek and silver-sided but only about eight inches long. She adjusts her physiology to fit a saline environment, seeks out prey she has never seen before, and learns to deal with tides and currents and avoid a new cadre of predators. Finding safe haven in eelgrass beds, she is already lucky to have made it this far; nearly all of her thousands of brothers and sisters have been eaten by birds, larger fish, or crabs. Against all odds, this young female survives and migrates up the Strait of Georgia, through the Discovery Islands and Johnston Strait, all the way to the really wild waters of Queen Charlotte Sound and even beyond, to Alaska. After two years spent feasting on ocean zooplankton, she swims all the way back south to the mouth of the Fraser River (see map 4). Heavy with her load of eggs, she begins the tough swim upriver, following the chemical signal of her birthplace eight hundred kilometers upstream. She averages about fifty kilometers per day.

Her first big hurdle is Hell's Gate, a narrow canyon of towering rock walls that forces the entire Fraser River through an opening only 115 feet wide. The name Hell's Gate was coined by explorer Simon Fraser, who in 1808 described this narrow passage as "a place where no human should venture, for surely these are the gates of Hell." Here, our sockeye swims against a strong current and massive cataracts before entering the smaller and more navigable Thompson River. Other salmon she has been swimming with stay in the Fraser itself and head for Stuart Lake. After about three weeks of swimming upstream, she reaches her destination and finds a bed of clean gravel, where she constructs a redd and lays her eggs, which are fertilized immediately by a waiting male. It is an end game she is playing, since once those eggs are safely tucked in the stream gravel, she dies. To this day, biologists do not fully understand the extremely sophisticated biology that allows salmon to navigate such long distances through dangerous and changing conditions.

Five different species of salmon make the journey to spawn in the upper reaches of the Fraser River system: Chinook (aka king,

Oncorhynchus tshawytscha), coho (aka silver, *O. kisutch*), sockeye (aka red, *O. nerka*), chum (aka dog, *O. keta*), and pink (aka humpie, *O. gorbuscha*). All these runs, except for the chum, have declined precipitously in the last decade. The numbers of Fraser sockeye have been steadily dwindling for an even longer period, since the mid-1990s. At that time, something began killing large numbers of returning sockeye—anywhere from 40% to 95% of fish in some years—before they could spawn. In 2009, more than 11 million sockeye were forecast to return but only 1.4 million showed up. The 2016 sockeye run was the lowest ever recorded and resulted in a closure of commercial and recreational fisheries. The extraordinarily low numbers observed in 2017 are unprecedented.

In one particular case, early investigations revealed very high mortality in the Cultus Lake runs of sockeye salmon. Cultus Lake should be an easy run for the salmon, since it's only a hundred kilometers upstream and well below the Hell's Gate cataract. From historic numbers of 70,000 fish, the population declined to less than 100 by 2004. Scientists concluded that this collapse was due to over-harvesting and warm El Niño conditions in the 1990s. The high mortality was also linked to a shift to earlier migration times; a team led by Simon Jones from the Pacific Biological Station determined that most of the early migraters were infected with a myxozoan parasite, *Parvicapsula minibicornis*. A multi-year captive breeding and research program was developed to figure out what was killing these salmon and to bring back the run. Although the role of the parasite remains enigmatic, as of 2012, there is now a relatively successful wild sockeye run to Cultus Lake.

The decline of Fraser River salmon is worrisome for many reasons. Wild salmon fisheries are central to the economy of British Columbia and vital to the cultural identity of First

Nations people. The aboriginal tribes of the Pacific Northwest have fished for salmon for all of their living and cultural memory. Respect and kinship is expressed in much of their art and is reinforced in tribal laws. Closer to home for me, the low salmon stock could mean starvation for our southern resident killer whale (orca) population. Depending on their age and size, each whale requires some thirty pounds of Chinook salmon per day, and the dwindling prey base is the biggest threat to their survival. J pod is the extended family of orcas that has frequented the San Juan Islands and Puget Sound for longer than my life span and is currently listed as endangered. Because orcas surface often and have huge, distinctively shaped dorsal fins and white saddle markings, I know every one of the twenty-three whales in our local pod by sight. Losing any one of them is a tragedy in our community, and the loss of even one breeding female tips the scales against the pod's survival. In years past, when Chinook levels were adequate, I could hear the whales feeding right off our beach several times a day and watch from my kayak as the entire pod—a close, multi-generation family group—swam by. This is the first year in many that I have not seen them here, because they are further up Georgia Strait and the outer coast of Vancouver Island, where they can find more food. I hope they are finding enough.

The Fraser River salmon face many challenges. There are multiple stresses in both river and ocean, with complex interactions. The salmon are plagued by parasites. The fresh waters of the river and the ocean waters they swim through contain pollutants that compromise their immune systems and disrupt their endocrine systems, making them more susceptible to the many different pathogens waiting to make a salmon their host. This is particularly true for the salmon returning to spawn—they are near the ends of their lives, and a pre-programmed dying process has already begun. Humans catch millions from the ocean just when they are reaching maturity. In recent years, they have faced another problem that has the potential to compound the others: high water temperatures. An unprecedented warm-water event in 2015—linked to a persistent area of high pressure in the Pacific called the Blob, which itself created an ocean heat wave that was first detected in 2013—was followed by a hot El Niño year in 2016. The Pacific Salmon Commission, a Canadian-American transboundary agency that helps manage fisheries, noted that the temperature of the Fraser River on August 18, 2016, at Qualark Creek, about twenty kilometers north of Hell's Gate, was 20.6°C, or 2.5°C higher than average for that date. Such warm river waters not only stoke disease, they can kill salmon outright from heat stress.

There are also major concerns that the growing salmon aquaculture industry could be harming wild salmon. Farmed salmon are extremely vulnerable to infectious disease, and it would be possible for an outbreak among diseased farmed salmon to spill over into wild populations. Additionally, some farmed salmon are infested with ectoparasites called sea lice (*Lepeophtheirus salmonis*), which can infect wild salmon swimming nearby. In the salmon aquaculture industry in British Columbia (as well as Washington State), Atlantic salmon are raised in net pens located in enclosed bays, often beside the routes that the wild salmon must traverse to return to their home rivers. This salmon farming is big business. In 2016, according to the British Columbia Seafood Industry the economic value of farmed Atlantic salmon in British Columbia was close to C$800 million.

After eighteen years of declines among wild salmon, a body called the Cohen Commission was established by Canada in 2009 to investigate the causes. It considered carefully the threat from aquaculture. After years of study, no clear evidence of harm from aquaculture operations was uncovered, but neither was salmon farming exonerated as a possible cause of the decline. So as wild salmon populations in the Pacific Northwest continue to diminish, it remains an open question whether salmon aquaculture is a contributing factor. At the same time, the threat of a new disease arising in farmed salmon and spilling over to infect wild salmon hangs over the native fish. To understand why scientists take these threats very seriously, we first need to take a global view of the salmon disease problem.

· · ·

In 1984, Norway led the world in the farming of Atlantic salmon (*Salmo salar*), raising them in very high-density conditions. That year, salmon in the pens started getting sick. The salmon grew pale, gasped at the surface, and became lethargic. Scientists discovered they were gasping because their red blood cells were damaged and not holding enough oxygen; they were severely anemic and hemorrhaging in several organs, including the kidneys. By the peak of the outbreak in Norway, more than eighty farms were affected, and mortality reached 90%.

Although an infectious agent was suspected, it took many years to identify and cultivate the agent. Since they could not be seen in early histological studies, bacteria and larger parasites were ruled out, leaving a virus as the probable culprit. Viruses are tough to isolate and identify because they require cell lines to be grown in pure culture for testing. Early attempts to isolate the virus in existing commercial fish cell lines were not successful. Finally, in 1995 Birgit Dannevig, Knut Falk, and their colleagues at Norway's National Veterinary Institute created a new cell line (SHK-1) from an Atlantic salmon kidney. In the late

1990s, the group concluded that the disease of farmed Atlantic salmon was caused by an orthomyxovirus. They named it infectious salmon anemia virus, or ISAV.

ISAV is a close relative of human influenza viruses. It is called an RNA virus because it uses RNA as its genetic material instead of DNA. DNA viruses are more common; they store genetic information in DNA and use RNA only as a messenger to transfer information from the DNA. DNA is a good genetic repository because it is stable. RNA is less stable. The genetic code in RNA is not checked or protected as carefully as the information coded in DNA, making it more subject to error when it is replicated to create new viruses. Errors in the genetic code are bad if you are a human, but if you are a virus trying to evade a host's immune system, errors can be beneficial because they allow rapid genetic change. RNA viruses can evolve much more rapidly than DNA viruses—more rapidly than any other organisms on earth, in fact—thus greatly increasing the chance of hitting on a new genetic combination able to get past the immune system defenses of a host. Consequently, RNA viruses have unusual power to outwit the immune system. Otherwise, the life cycle of an RNA virus is similar to that of a DNA virus. It attaches to and penetrates a host cell—in this case, the red blood cell of a fish. Then it hijacks the reproductive machinery of the cell to create many, many copies of itself and sends these new viruses out in protective capsids to infect other host cells. The gills of an infected fish look pale because its red blood cells are compromised and cannot hold enough oxygen.

ISAV made history when it was transported from Norway, causing massive outbreaks and shutting down salmon farms in other parts of the world. In 1996, a new disease of farmed salmon in eastern Canada, hemorrhagic kidney syndrome, was verified as also being caused by ISAV. ISAV was detected in Scotland in 1998 and thereafter in the Shetland Islands. By 2000 it had spread to the Faroe Islands and in 2001 to Maine. In another outbreak in 2012, ISAV hit farms in Nova Scotia; five months later it spread to Newfoundland and Labrador, causing 750,000 farmed Atlantic salmon to die. In Chile, ISAV was initially detected in 1990 and 2001, but it wasn't until June 2007 that a massive outbreak started there in farmed Atlantic salmon. ISAV was also isolated from farmed coho salmon in Chile.

Fortunately, fisheries managers have developed successful control measures for ISAV. The general strategy is one of increased hygiene and stringent culling and isolation procedures. It's called "all in, all out." Each production site is limited to a single generation of fish. Between generations, all equipment must be dismantled, thoroughly cleaned, and inspected by a veterinary commission. There is careful monitoring for ISAV, and when it is detected, the entire stock is immediately culled. While best practices like "all in, all out" were being established to combat the outbreak in Norway and Scotland, Chile's rising salmon farming industry was initially more lax and was rewarded by the disastrous outbreak of ISAV in 2007. By 2010 the Chilean industry had lost half its farmed fish and $2 billion in revenue. Chile's pathogen hygiene protocols improved in the wake of this debacle, and production of farmed salmon in Chile is now second only to that in Norway. In fact, salmon is Chile's second largest export, just behind copper.

In addition to maintaining stringent hygiene practices, fish can be vaccinated against viruses. Vaccines are preparations of antigens derived from pathogenic viruses that are rendered non-pathogenic. They stimulate the immune system and increase resistance to disease from subsequent infection by the specific pathogen. Fish can be vaccinated individually by injection or by adding the vaccine to their feed. ISAV vaccines have been added to salmon feed in most places for over a decade. As with most vaccines, the ISAV vaccines do not offer complete protection in Atlantic salmon, but do increase resistance. By combining vaccination with the "all in, all out" hygiene policy, aquaculture farms have so far been able to prevent outbreaks.

Whether on land or the sea, it is challenging to nail down details about how an infection spreads from host to host. Disease in the ocean is particularly tricky; it is hard to follow salmon around, and many infectious agents can live freely in seawater. Easy transmission is one reason ISAV was such a problem. ISAV, which can live for up to eight days in seawater, spreads easily through the water and can also be vectored from fish to fish by the ever-present salmon lice. Scientists in both Maine and Chile have also shown that ISAV can be transmitted through seawater in shipments of live fish and occasionally via sea lice and contact with infected wild salmonids. This direct transmission in seawater complicates control measures and allows potentially explosive outbreaks.

In the aftermath of the slow-rolling global outbreak of ISAV, all eyes are on the richest salmon bounty in the world: the five species of wild Pacific salmon in Oregon, Washington, British Columbia, and Alaska. This region is not only home to the greatest runs of wild salmon, but also to a massive salmon aquaculture industry, especially in British Columbia. One of the BC salmon farming regions is north of the Fraser River, where Atlantic salmon pens fill the tight passages of the Discovery Islands (see map 5). These are the very waters that the wild Pacific salmon must migrate through on their way to rich feeding areas north of Vancouver Island. It's one reason there is so much concern about salmon farming in British Columbia and such strong suspicion that ISAV or other salmon viruses might

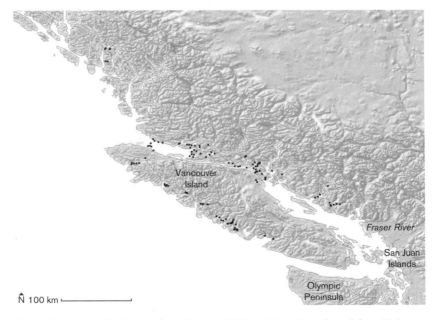

Map 5. Location of Atlantic salmon farms in British Columbia. Adapted from Living Oceans, map of BC Salmon Farms.

be playing a role in the decline of the Fraser River wild salmon runs.

Is ISAV present in the Pacific Northwest? This seemingly basic question has been challenging to answer. It has generated heated controversy and legal actions. There were reports in 2009 that ISAV had been detected in wild British Columbia salmon. These reports were not readily confirmed by the government. The delay generated significant public protests and headlines suggesting that the government was covering up the disease threat to protect the lucrative aquaculture industry. It turned out that the main science laboratory that tests worldwide for ISAV, the Kibenge Lab, used unapproved molecular primers in its testing and likely detected a non-pathogenic but related virus. The lab lost its international accreditation in 2012. Official surveillance by the Canadian Food Inspection Agency of thousands of wild and cultured fish has continued to find no ISAV in British Columbia, and there is no detection of ISAV in salmon from US waters.

The controversy about whether there is a danger of spillover from farmed Atlantic salmon to wild Pacific salmon may have some legitimate roots in the complicated biology of pathogenic viruses and the challenges of detecting pathogenic viruses in the wild. A new study by the Kibenge Lab, published in 2016, claimed that 9 out of 709 wild fish tested positive for a partial genetic sequence of ISAV. The lab reported new segments of the virus genes that are not reliably detected by existing diagnostic tests. If the virus is present but its genetics are changing, the emergence of new genetic variants could explain why its detection has been difficult. A constant challenge in the ecology of infectious disease is the rate at which new genetic strains emerge, and the lottery-like unpredictability of when or how a new very virulent strain will take hold.

As a result of this ongoing controversy in Canada, there are enhanced protocols in place for detecting ISAV in US West Coast waters. Since new dangerous strains can rapidly emerge, the current stringent surveillance protocols are essential.

· · ·

A few tantalizing clues may help answer the question of what is reducing the numbers of Fraser River salmon. Despite the overall trend of decline, in one of the tributaries close to the mouth of the Fraser, the Harrison River, salmon are thriving and not declining. The sockeye salmon from the Harrison River are a distinct genetic subpopulation and have very different migration patterns.

Instead of living in freshwater lakes for a year, they enter the ocean sooner and take a different route to get to rich, pristine northern waters. They migrate down past the San Juan Islands, out the Strait of Juan de Fuca, and up around the west side of Vancouver Island (see map 4). This means they never have to pass through the waters on the more populated east side of Vancouver Island or the tight passage of Johnstone Strait with all its salmon farms. These differences suggest three explanations to me for why this run is doing better. It's possible that entering the ocean sooner may be advantageous in today's ocean. The migration route to the west of Vancouver Island may be less stressful for the fish because it has more food, less acidic waters, or less fishing pressure. Finally, this alternate route may have lower levels of pathogenic microorganisms or sea lice.

Although Pacific salmon seem for now to be unaffected by ISAV, there are other infectious diseases taking a large toll on their populations, and these may partially explain the Fraser River declines. A recently emerged disease of Chinook salmon in the Yukon River in Alaska is caused by the protistan *Ichthyophonus hoferi*. Although its levels have not been studied in the Fraser River, it is present in Salish Sea waters. This protist is a fish-killing pathogen with a broad host range. It has been identified as the cause of disease in over a hundred species of marine, freshwater, and brackish-water fish since its original classification in 1911. Infection is debilitating and can be lethal in many species, including herring, salmon, and various flounders. Research led by Carl Huntsberger from the Coonamesset Farm Foundation showed that outbreaks of this disease occurred recently in herring from the northwest Atlantic, with the disease remaining at low, endemic levels between outbreak events. Although all the routes of transmission in the wild are not yet known, it is bad news that it can be transmitted both directly from fish to fish and indirectly through predation. In the laboratory, salmon become infected by eating infected Pacific herring. Laboratory-reared Atlantic herring (*Clupea harengus*) and mummichogs (*Fundulus het-eroclitus*) can be infected by eating food contaminated with *Ich-thyophonus* spores and develop infected organs and skin lesions within 18 days, with death occurring between 70 and 110 days. If it breaks out, this will be a wicked disease to manage because it infects a large range of species: in addition to infecting herring, salmon, and halibut on the West Coast, it can maintain low-intensity infections in surf smelt (*Hypomesus pretiosus*), Puget Sound rockfish (*Sebastes emphaeus*), Pacific tomcod (*Microgadus proximus*), and speckled sanddab (*Citharichthys stigmaeus*).

Besides killing fish directly, *Ichthyophonus* can threaten preda-tory fish like salmon indirectly, by killing their prey. Herring (*Clu-pea pallasii*) are one of the most important forage fish in the North Pacific, the mainstay food of all the salmon as well as seabirds and even seals. Herring are highly susceptible to *Ichthyophonus* infec-tion. From Puget Sound to the Gulf of Alaska, herring are infected by *Ichthyophonus* at a rate of 30 to 40%. *Ichthyophonus* preferentially infects older fish, which happen to be those with the greatest reproductive potential. Paul Hershberger, one of the members of our Research Coordination Network for Ocean Health (along with Carolyn Friedman, Kevin Lafferty, Colleen Burge, Joleah Lamb, and forty other scientists), is one of the leading fish health experts for the US Geological Survey in Seattle. He has been studying *Ichthyophonus* in fishes from the North Pacific for two decades. Although it is hard to document how many herring are dying from *Ichthyophonus* infection or what impact it is having on population levels in such a wide-ranging species, Paul's group suggested in a recent publication that a decrease in the mean age of adult Pacific herring in Puget Sound is caused by high mortality from *Ichthyophonus*. Early studies show that it may be taking a large toll on herring, and this could have disastrous effects on the larger predatory fish, like salmon, that depend on herring.

A link between *Ichthyophonus* outbreaks and climate change is likely but unproven, owing largely to the difficulty of quantifying disease impacts in wild marine fishes. In the laboratory, higher temperatures have been shown to increase the severity of ichthyophoniasis, suggesting that *Ichthyophonus* infections could increase in warming oceans. In *Ichthyophonus*-exposed rainbow trout, increasing temperature makes all impacts of the disease worse: infection prevalence increases, cumulative mortality increases, and host swimming performance declines. To show a link with warming in nature, Richard Kocan, a colleague of Paul Hershberger's at the US Geological Survey, made field observations during the Yukon River outbreak in Chinook salmon that revealed elevated river temperatures likely contributed to disease progression.

Yet another serious disease of salmon in the Pacific Northwest is infectious haematopoietic necrosis virus (IHNV). IHNV is an RNA virus like ISAV. One of its closest relatives on land is the rabies virus. IHNV is described by the World Organization for Animal Health as a virus that began endemic to the Pacific Northwest, and then spread in infected salmon eggs to Asia and Europe. In the wild, it is mostly considered a disease transmitted and spread in the freshwater river phase of a salmon's life, and it can devastate juvenile sockeye salmon. It turns out that Atlantic salmon are more susceptible to IHNV and shed more virus than wild Pacific salmon when infected, so they are also a potential reservoir for the virus. Of the three outbreaks that have occurred so far in farmed BC salmon, the 2001–3 outbreak is notable in its spread among farms. It began with a single farm in the Discovery Islands. Genetic sequencing by Sonja Saksida of the British Columbia Centre of Aquatic Health Sciences traced the virus's spread to thirty-two salmon farms across the southern half of the British Columbia coastline—81% of the farms in the region. Wild sockeye are highly susceptible to IHNV. Some of the most frequent and best-recorded outbreaks have occurred in wild sockeye and salmonid enhancement facilities.

.   .   .

Atlantic salmon have adapted extremely well to being farmed, and business is booming. It is easy to spawn salmon eggs in hatcheries, raise them in small tanks to one month old, and then deliver the young to huge net pens in the ocean, where they are given feed made from small fish and grow rapidly. Breeders have worked magic and developed salmon varieties that reach maturity in one year and are more disease resistant. In two years, delicious, healthy salmon can be raised from egg to table. The Food and Agriculture Organization of the United Nations reports that production of farmed Atlantic salmon increased from 28,000 tons in 1984 to 2.3 million tons in 2014. The salmon farms of Chile, growing a variant of the Atlantic salmon, produce almost as much salmon as all the wild runs combined. Given the global demand for fish and marine protein and the decline of wild fisheries all over the globe, it seems clear that fish farming is a vital part of the solution to the problem of how to feed a growing human population.

However, the devastation caused by the ISAV and IHN viruses and the parasite *Ichthyophonus* in both wild and farmed salmon must certainly give us pause. Salmon farming is less viable if penned salmon remain vulnerable to multiple diseases and act as reservoirs for pathogens that impact wild fish. Thus we must develop even more stringent protocols, best practices, and careful surveillance methods to safeguard the world's aquaculture industry *and* wild fish populations.

It is also important to guard against threats other than disease that farmed Atlantic salmon pose to our wild Pacific salmon. When farmed salmon escape, they compete for food with wild salmon, might interbreed, and also share diseases. The danger of escape is ever-present, as we found out in the fall of 2017 in the San Juan Islands. There is a big Atlantic salmon farm to the east of San Juan Island on Cypress Island. I learned about this farm when its nets failed. As many as 300,000 two-year-old, ten-pound Atlantic salmon made a successful break for the wild. The farm managers initially blamed the solar eclipse and unusual high tides for compromising their pens, but oceanographers immediately discredited this story by reporting the absolute lack of anything unusual in the tides or weather. It was faulty engineering and deferred maintenance that allowed the massive release of non-native fish. The community was outraged by both the escape and the attempt to evade blame. A similar spill had occurred on this same farm twenty years earlier. This time, the fish and wildlife department immediately called on all fishermen to rush out and catch as many escapees as possible. It was fun for some of the locals, but the gill-netters who pulled in hundreds in a haul could sell them for only a dollar per pound. Some of the fish strayed as far as the west coast of Vancouver Island, and a few made it into important wild salmon spawning streams, including the nearby Skagit River. The release left us all feeling shocked that there was no viable plan for dealing with an accident of this magnitude. A major fish spill may not be as bad as an oil spill, but both require better mitigation plans than currently exist in our waters.

Sometimes these problems go on for years without a policy response, but not this time. A mere eight months after this Atlantic salmon spill, Governor Jay Inslee banned all Atlantic salmon farming in Washington State. His legislation will prevent new farms and will phase out all existing farming by 2025, bringing to an end three decades of non-native fish farming. Washington State now joins Alaska in banning commercial finfish aquaculture. The decision is controversial and pits the economic interests of a large Canadian corporation against concern for the environment and Washington's identity as a provider of wild salmon. "We have done the right thing," Senator Kevin Ranker said. "We have supported our culture and our natural resources." It remains to be seen what the impact of this historic, controversial legislation will be on the large Atlantic farming enterprise in nearby British Columbia.

I recently visited the Hakai Institute Laboratory on Quadra Island, just south of the Discovery Islands fish farms in Johnstone Strait. This is not an easy place to get to. The CEO and founder of the Hakai Institute, Eric Peterson, had invited me and colleagues from Friday Harbor Marine Labs to visit as part of our new collaboration to study transboundary disease outbreaks in eelgrass. I boarded a seaplane with Eric and colleagues in Friday Harbor and flew north for a stop in Nanaimo, British Columbia. We cleared customs in Nanaimo and then headed north again to Quadra Island, wending our way up the Inside Passage, covering the same route from the air that our salmon would swim below in the sea. Along the way, we got stuck in a fog bank, which forced us to backtrack for another thirty minutes to find clear visibility. About two hours after leaving gray and overcast Friday Harbor, we coasted in for a landing on a sunny day on Quadra.

Both of the Hakai Institute research centers, the one on Quadra Island and the even more remote one further north on Calvert Island, are like marine science fantasy camps for me. Eric Peterson is an insightful visionary who cares deeply about marine biodiversity and has transformed our capacity for marine conservation in these cold, high-biodiversity waters of coastal British Columbia. He has single-handedly created a research institute in a region that needs to be studied in this time of a highly impacted, changing ocean. He has equipped it with a crack team of dedicated staff scientists and state-of-the-art instrumentation, and it is run with around-the-clock efficiency. It is a vital listening post on a changing ocean. Our oceans would be in much better shape if more funds were invested in this kind of research infrastructure.

One of the institute's projects focuses on salmon health. I was pleased to tour a facility that is the result of a very intensive effort among scientists from Hakai and the British Columbia government to regularly sample both the virus load and the physiology of thousands of young wild salmon before and after they run the gauntlet of fish farms. This level of surveillance will reveal any harm done to the wild salmon by the fish farms. It is a good example of the innovations and partnerships that increase as aquaculture continues to grow as a source of ocean-derived protein.

When considering the impact of disease on commercially valuable marine species, salmon evoke some of the greatest concern because of the high value and large size of both the wild fisheries and salmon aquaculture, and because of the well-documented history of outbreaks. But salmon are far from the only commercially harvested animals in the ocean that have suffered from disease outbreaks and remain vulnerable to new epidemics and debilitating low-intensity infection. Three years ago, our Ocean Health team met to investigate the available information on economic losses due to marine diseases. Our team, led by Kevin Lafferty, revealed that infectious disease hammers ocean-based foods with a price tag in the billions of dollars. There are sixty-seven diseases that lead to mortality and have proven to be costly. We tallied examples of diseases in which economic losses could be pinpointed; of these, 49% affected fishes, 21% crustaceans, 28% molluscs, and 1% echinoderms. Twenty-five percent of the identified infectious agents are viruses, 34% are bacteria, 19% are protists, and 18% are parasitic metazoans (multi-celled animals such as worms). Their impacts are greatest in the crowded conditions of aquaculture, which increasingly dominates seafood production in the face of declining wild fisheries. Marine diseases of farmed oysters, shrimp, abalone, and fishes—particularly Atlantic salmon—cause billions of dollars of losses each year. On the positive side, big advances are being made in breeding disease-resistant stocks and developing new vaccines. In comparison, it is difficult to estimate disease impacts on wild populations in the open ocean. We have little data on how often farmed species receive infectious diseases from wild species and, in turn, incubate and convey infectious agents back to wild species.

Infectious disease in the ocean is a thief of enormous proportions. It's such a good thief that we often don't realize we are being robbed. We need to up our game of controlling disease in wild and farmed species like salmon, and there is a lot more we can be doing, as I discuss in the last chapter. One thing is certain: it would be an immeasurable loss if salmon runs ceased to surge through our few remaining wild rivers and it were no longer possible to taste the rich flesh of a salmon that had spent most of its life swimming freely in the ocean, living the life that nature intended.

CHAPTER FIVE

Starfish Outbreaks

An Ecological Domino Effect

If it is a terrifying thought that life is at the mercy of
the multiplication of the minute bodies, it is a
consoling hope that science will not always remain
powerless before such enemies.

    Louis Pasteur

As a brand-new graduate student at the University of Washington in 1983, I traveled by fishing boat with Bob Paine to Tatoosh
Island—an uninhabited, wave-swept speck of rock half a mile
seaward of Cape Flattery, off Washington's Olympic Coast. Everything about Bob was larger than life. At six feet four inches, he
towered over me. His voice could modulate from the surprisingly soft tone he used to express his wonder at Tatoosh and its
critters to the booming tone he used when he bragged about
himself or one of his students. He was known to many as the
most brilliant field ecologist of his time. For Bob, Tatoosh was a
source of inspiration. Its rocky intertidal was his touchstone, the
backdrop that allowed him to see bigger patterns and processes
in nature. As we approached Tatoosh in the early morning, huge

waves crashed on steep cliffs underlaid by broad, algae-covered, rocky benches. I like adventure, but I looked on nervously as we approached an increasingly rough, misty shore. How would we ever land? At the last minute, we tucked into a protected bay, scrambled into the small rubber Zodiac safety boat, and rapidly offloaded, commando style, on a rocky beach.

We got to work at dawn the next morning to catch the low tide. Bob and I and three other students followed a rough trail from the lighthouse cabin down to the intertidal and across a cobble beach to the exposed rocky benches of Tatoosh. In the morning light, they were covered in bands of different textures with all the riches and colors of the sea: bright red and brown and green bands of turfy algae, shiny black mussels as big as squirrels, bright purple and orange starfish, green anemones the size of dinner plates, and green sponges carpeting the rock underneath. I walked over a mussel bed and Bob pointed across a surge channel to the section of rocky intertidal where, twenty years earlier, he had removed every single ochre starfish. This was his now classic experiment to reveal how the biota of the intertidal changes when the top predator is removed. He had taken all these starfish and released them into another section of intertidal further down the shore, because he also wanted to see what would happen if he increased their numbers. Removing starfish from the intertidal may sound simple, but the Tatoosh intertidal is a dangerous place, battered with twenty-foot waves. As we watched, Bob said, "Here comes a rogue wave, step up further." A big wave hit and roared forty feet up shore, right to where we had been standing, then sucked back to reveal forty feet of the subtidal.

Following Bob's experimental manipulation, the stretch of shore without ochre stars became encrusted with what Bob called a "mussel glacier"—a huge dark mass of thousands of mussels packed together stretching from the subtidal to far up in the intertidal. The mussel bed crowded out many other species, like green sea anemones, green sponges. and pinkish sea squirts, that would otherwise live there. Around the corner, where he had added the ochre stars, the existing mussel bed was pushed far down into the subtidal, and the shore was alternately dominated by green sea anemones, bright pink coralline algae, spiny purple sea urchins, and a garden of multicolored algae. These results showed that the ochre star controlled the fates of the species around it. The star was, in Bob's mind, very much like the center stone at the top of an arch that holds up all the other stones and, if removed, would cause the whole structure to collapse. So Bob coined the term *keystone species* in a 1969 note about food webs for the kind of ecological role the starfish assumed in the rocky intertidal.

Because of its elegance, clear-cut results, and broad application in ecology, Bob's experiment with ochre stars made history and is now considered one of the greatest experiments in nature. It produced not only the ecological concept of keystone species, but also that of trophic cascade, which describes the changes that reverberate down a food chain when a species at a higher position, or trophic level, experiences a change in its population. During the entire time I prowled the island's shores with Bob, I remained aware that several key processes that occur in habitats everywhere, in the ocean and on land, were described first here in the Tatoosh intertidal.

When I first heard the news about starfish dying on the West Coast of the United States in late August 2013, Tatoosh and its lessons came immediately to mind. Within a few months, it would become clear that nature was implementing an experiment in keystone species removal on a truly epic scale.

. . .

On September 6, 2013, I received a pile of disconcerting emails. Chris Harley, a professor at the University of British Columbia (and a former student of Bob Paine's) was forwarding information about a big outbreak of disease in sunflower stars in British Columbia. On the blog of echinoderm expert Chris Mah, Echinoblog, people were posting striking photos, and I watched a video made by Jonathan Martin, then a research associate at Simon Fraser University in British Columbia. Martin posted this description with his video:

> I just got back from a dive out in West Vancouver, and there seems to be a huge mortality event of some kind.... The animals seem to waste away, "deflate" a little, and then just ... disintegrate. The arms just detach, and the central disc falls apart. It seems to happen rapidly, and [it is] not just dead animals undergoing decomposition, as I observed single arms clinging to the rock faces, tube feet still moving, with the skin split, gills flapping in the current. I've seen single animals in the past looking like this, and during the first dive this morning I thought it might be crabbers chopping them up and tossing them off the rocks. Then we did our second dive in an area closed to fishing.... The bottom from about twenty to fifty feet was absolutely littered with arms, oral discs, tube feet, gonads and gills, to the extent where it was kind of creepy.

Chris Mah speculated that we might be observing the results of a famine caused by overpopulation or disease. Others chipped in that maybe it was caused by low salinity shock from the runoff produced in the recent rains, or low oxygen or ocean acidification. Questions were raised: Was it related to a similar decline of East Coast stars? Was the incident restricted to British Columbia? Melissa Miner, coordinator of the Multi-Agency Rocky Intertidal Network (aka MARINe) conveyed that there were earlier reports from Washington State of sick *Pisaster,* the intertidal ochre star.

I followed this carefully because I run the Research Coordination Network for Ocean Health, a research coordination network funded by the National Science Foundation. The many scientists in the network form a detective team whose job it is to solve mysteries of infectious disease outbreaks in marine organisms, from corals to oysters to abalone to urchins to dolphins. Should I activate the network so that we might collectively get to the bottom of what was going on with the starfish? While every mass mortality event seems unique, there are common themes and it helps to have a team of experts to share strategies. In the preceding few years, I had dealt with two starfish mortality events that had proven frustrating. While running a training course at Friday Harbor Labs for doctoral students as part of our network for marine health, there had been a small die-off in our waters of the intertidal ochre star, and we had tried to get samples as a teachable project for the class. That event was over so fast we failed to get enough samples. Before that, we had received samples of sick stars from Alaska, but pathologists at Cornell's vet school concluded that the cause was not an infectious agent, but rather heat stress. Having experienced these two failures to understand the role of infectious disease in starfish mortalities, I was determined that we would do our best now to figure out what was killing these keystone stars.

Before the day was out, I sent my colleague at Cornell, Ian Hewson, an email asking if he would look at samples from dying stars from British Columbia. Ian's research specialty is genomics, with a focus on viral genomes. He is one of a rare breed of scientists trained to read the labyrinthine genetic code of viruses. He had been working with non-pathogenic viruses of Hawaiian sea urchins—distant cousins to starfish, in the same phylum, Echinodermata—and was probably the only scientist in the world at the time with experience working with echinoderm viruses. Ian replied that he was excited about investigating a new disease, especially if it involved starfish. "Count me in if there is mass mortality," he wrote. In a series of exchanges, we worked out how we would get our samples to Cornell. By the end of the day we had a plan: I would send or request samples that had been flash-frozen to −80°C, and Ian would test them for both bacteria and viruses. I also contacted Harley Newton, a pathologist from the Bronx Zoo, who agreed to create histological slides of samples preserved in formalin. It didn't get better than this—we had the world's experts lined up to examine these diseased animals. Harley's histology would allow the detection of anything bigger than 0.5 microns, such as larger bacteria, protozoans, fungi, and parasites. Ian's metagenomics would reveal the presence of genetic material from smaller, potentially pathogenic agents such as small bacteria and viruses.

We tried to get the samples from Canada, but ran into some hassles and delays with shipping sick critters across the border. Three days later, on September 11, I heard of Steve Fradkin's observations that the ochre star, Bob Paine's original keystone, was also sick, and he was efficient in collecting and sending the kind of samples we needed. Steve works for the National Park Service and was responsible for monitoring starfish in the pristine and wild waters of Washington's Olympic National Park. At the same time, Melissa Miner spread word of the outbreak to her intertidal network, which extended from San Diego to Alaska. By September 15 we were hearing of other small events affecting intertidal ochre stars in California, and we arranged to have samples from some of those sick stars shipped to Ian. On

October 9, I received a surprising email from Marty Haulena, a veterinarian from the Vancouver Aquarium, reporting that stars in their captive collection were dying. It seemed to be turning into a more widespread outbreak than we had expected. I was scheduled to go to Indonesia for a month on October 21 to lead a coral health project. I was imagining that this mortality event would be similar to earlier starfish die-offs and would end by the time I returned, as the waters cooled in advance of winter. Although I was betting on infectious disease as the cause of the die-off, there was still a high degree of uncertainty. We live in strange times, with a rapidly changing ocean, so we couldn't discount some other cause like pollution or a sudden jump in the progress of ocean acidification.

By early October, Ian and Harley had received some samples from sick stars, but Ian in particular needed many more. There would be thousands of viruses and bacteria associated with each starfish sample that he sequenced. But if he ran enough samples, any causative pathogen would, he hoped, eventually stand out in the results. Field ecologists were sending samples directly to Ian, and we had some of our own that Joe Gaydos, director of the SeaDocs Society, had collected while doing a hundred-transect survey in the San Juan Islands. Joe is an intrepid diver as well as a talented veterinarian, and he would be an important ally in the long work ahead. I was too busy with other projects and planning my imminent trip to Indonesia to coordinate future samples. Fortunately, I had the perfect person at hand to take over—Morgan Eisenlord, a lab technician already working at Friday Harbor Labs. On the verge of applying to PhD programs in marine ecology, Morgan had come to take marine biology courses at the labs. She was excited to be involved in any kind of project that would teach her new skills.

Morgan and I met in my laboratory in the Gear Locker to go over dissection protocols with some sick starfish. The Gear Locker is a small lab but a great work space, with a big work table and windows overlooking the bay and the town of Friday Harbor. The center of the lab is filled with running saltwater tanks; at that time they housed the stars that Joe Gaydos had recently brought in for us to examine. They didn't look too bad. One of the mottled stars seemed a bit low in turgor and was missing one arm, but it didn't have any lesions or clear signs of wasting. We bleached our entire work surface, bleached and ethanol-fired our surgical tools, and donned plastic gloves. As Morgan watched, I cut into the top of the star and removed a three-inch by three-inch window of surface tissue, thick and stiff with embedded skeletal spicules. This exposed the stomach and hepatic caeca—also known as the liver—in five-fold symmetry, since there was a projection of the stomach and the caecum into each arm. In the center of the star and underneath the other organs was a second stomach, called the cardiac stomach, directly above the mouth.

I carefully removed a piece of caecum, then a section of pyloric stomach, and then the underlying cardiac stomach, dipping my scissors and forceps in alcohol and holding them in a flame between each step. Morgan labeled a plastic ziplock bag for each organ and then immediately took the fresh samples to the −80°C freezer. Meanwhile I followed pathologist Harley Newton's request, separating an entire arm from the body, cutting cross sections, and placing these in a formalin preservative. Harley would make slides from these cross sections and then examine them under the microscope. Because I cut each sample from the arm, all the organs were contained in each cross section. In a single slide, Harley could look inside the digestive caecum, pyloric stomach, tube feet, and surface skin of the star. The ultra-frozen samples would be sent to Ian Hewson, who would sequence them completely, revealing any genetic code that was present and thus the identities of any bacteria and viruses.

I turned the next dissection over to Morgan so that she could practice her methods. She unwrapped her own dissecting tools, sterilized them and cut into the dorsal surface of another star with a steady hand. She was already well trained in basic starfish anatomy from her courses and had no trouble repeating the entire protocol, making her own suggestions for improvement as she worked through the dissection. We wrapped up and I knew I could trust her with dissecting the rest of Joe's early San Juan Islands samples, and with the subsequent samples that were being sent to us from the outer coast of Washington and California. Morgan would end up processing over sixty animals while I was in Indonesia and sending them on to Ian and Harley. These included both symptomatic and asymptomatic stars; by comparing them, Ian and Harley could identify the differences in the sick stars.

We pinned most of our hopes on Ian's genomics. There is no hiding from an analysis that reads the sequence of all the DNA present. Screening for viruses in starfish was exciting for other reasons, too. Few viral infections of non-commercial species had been identified or studied. Partly because it requires highly specialized training to study the tiny, nearly unseeable viruses, there are very few marine virologists, and little is known about the biology of these smallest, yet most abundant organisms in our oceans. Really, how lucky was it that Ian was actively studying the viruses of starfish relatives already? It was an unusual chance to do novel science and advance our field.

Ian, a tall, blonde Australian and with a rough outdoorsman look, was a relatively young microbiologist in 2013, and still an untenured assistant professor at Cornell. Underneath his casual, approachable style is a genomics whiz kid. In his work, he grinds up an entire organism, or even the thousands of organisms preserved in a plankton tow, and uses the latest innovations in molecular biology to read the sequence of every gene in the sample. He is trained as a virologist, so his special focus is the genomics of viruses. In 2013, Ian was one of only a handful of ocean virus experts. He was probably the only person in the world who had previously worked with viruses in starfish relatives. His lab was fully equipped with the instruments needed for extracting and sequencing virus DNA, and he was adept at the gargantuan job of programming computers to interpret the millions of lines of genetic data that the sequencer would spew out.

Hundreds of ultra-frozen samples from both lesioned, sick stars and healthy-appearing stars were arriving at Ian's lab at Cornell. He organized his Cornell undergraduate students and hard-working technicians like Jason Button into an assembly line to process the samples. After the DNA was extracted, the samples were run on a genetic sequencer. The lab would run around twenty samples at a time, with each run taking a week to complete. The midnight oil was burning constantly in Ian's lab during October, November, and early December. The amount of work they did is all the more impressive when we remember that this was a sudden event that completely reoriented all the priorities already under way in his lab and his life.

Although we already knew this was a big outbreak by the time I received Pete Raimondi's report from Monterey at the December Nature Conservancy meeting, Pete's email changed my view of the outbreak. I had been viewing it in relation to previous outbreaks. From this perspective, it was fascinating but relatively normal, and we were following up to study it to better understand an ongoing infectious disease outbreak in an ecologically important species. Pete's description of the extremely rapid death of over ten species in one place, far from the original outbreak sites, shattered my sanguine framing. I suddenly realized we were facing a potentially new infectious agent that could be catastrophic for starfish biodiversity, and I needed to get involved in my own fieldwork as soon as possible. I checked my tide chart and phoned the airline, trying to get back to Seattle the next afternoon.

.   .   .

The day after I received the calls from Laura James and Katie Campbell at the meeting in California, I boarded a plane bound for Seattle along with my oceanographer husband Chuck. We stepped off our flight in Seattle at 4:00 pm and ran for our car, hoping to beat the rising tide at Alki Beach. As we drove through the darkening, cold, wet streets of wintertime Seattle, stopping to buy flashlights, warm hats, and plastic gloves, I shook my head and told Chuck I had a really bad feeling about this. I realized that my year was about to change. We talked about how it looked to be a huge hit to *Pisaster,* the keystone species, and how it seemed to be affecting all the starfish. "What if it drives some of these species to extinction?" I asked rhetorically. "All of our West Coast stars are endemic; they live nowhere else on the planet." I told Chuck I was scared and worried that there was no money to do the huge amount of work that needed to be done right away.

We reached the rain-slicked parking lot near Cove 1 at Alki Beach and found Laura in a nearby restaurant, having dinner with her dive buddy and working on her iPad. Meeting this small, dark-haired woman with a big smile for the first time, you would never guess she is a tough, intrepid diver and talented videographer. Going out night after night in the cold dark waters around Seattle, she did things like track giant Pacific octopus and film their babies as they hatched. Laura and a fleet of citizen divers, including Jan Kocian from Whidbey Island, would soon become my heroes for keeping an eye out underwater and educating the public about the health of our ocean biodiversity. They would be with us during the entire epidemic, sending weekly reports. Laura and her dive buddy were busy designing a clever app that would allow any person to snap a picture of a sick star and send it to her site, where it would arrive geo-referenced and help produce a map of sick star locations. We would later use some of the reports from Laura's app to track locations for collecting stars.

We put on our headlamps, donned plastic gloves, and pulled out our notebooks. Clambering in the dark over slippery boulders, we move down the beach toward the receding tide. Our headlamps picked up looming boulders and the edges of pilings on the dark shore as we crunched across gravel, looking for starfish bodies. Across Elliott Bay we could see the bright lights of Seattle. "Look, here's an arm," Laura called. "Yeah, that's an *Evasterias*," I replied as I hunkered down to look at the deflated, ripped-off arm, tube feet moving slowly. "Oh, and I guess that's the rest of it," I said as I pointed to a flattened, almost-dead four-armed star. We saw four purple stars clustered together, half in the water, with twisted arms and small lesions; we would come to recognize these as tell-tale signs of the wasting disease. As we walked along the water's edge, we found more and more arms and parts of stars. I realized, with a sick feeling, that the beach was littered with them. Many were clearly dead, but occasionally, we would find one with moving tube feet, or a star in the process of having its arm tear away from the body and walk off on its own. It was so unreal and unnerving. None of us, except Laura in her earlier dive, had ever seen anything like this. It was a macabre scene, but not all the stars were sick or dead. Chuck wandered higher in the intertidal and turned over a huge, flat slab rock. Underneath he could see in his light a cluster of seven purple stars, all looking healthy. "Here, these ochre stars all look fine," he reported. "They are OK."

Wanting to do a quantitative, more scientific accounting of the situation, I divided the beach into three regions of roughly equal size, and we took counts of dead and dying adult stars in each region. Of the 130 ochre stars we counted in one of these sections, about 65 did not show outward signs of disease. Expecting only dead and dying stars from Laura's report of the underwater carnage from the previous month, we were surprised to find any asymptomatic stars at all, let alone half the stars on the beach. In addition, it appeared that position on the beach was linked to mortality. Almost all the stars in the subtidal had died; they were dying in high numbers at the water's edge; but they seemed healthy in the higher intertidal (see figure 10 for a photo of dying stars). However, the mottled stars were more abundant than the ochre stars lower on the shore, and there was a chance that they were the more susceptible species. Our sample size was too small to tell. Thus, what looked like a pattern linked to tidal height could really be a pattern driven by the different distributions of the two star species. In any case, the differential mortality was a surprising new piece to the puzzle. At this site, why were the deeper-water populations suffering more? Perhaps these stars had more exposure to the infectious agent because they were underwater for longer periods. Or maybe drying out or being exposed to sunlight during low-tide periods inactivated the causative microbe.

Figure 10. Mortality of the ochre star in Bamfield, British Columbia. Photo by Drew Harvell.

My brain was in overdrive trying to detect every clue to the cause and any patterns driven by variation in the environment. Was there any new evidence to support my hypothesis that the die-off was caused by an infectious, transmissible agent, as opposed to a low-oxygen or strange pollution event? The unexpected better survival in the intertidal at a site where all the subtidal stars of the same species were dead was a big clue, but what did it tell us about why the underwater zone was so lethal? Were the intertidal stars escaping a new lethal pollutant carried down the Duwamish River—which was on the edge of an EPA Superfund site—or was the freshwater a route carrying in a new infectious agent from land? If an infectious agent was the cause, why was it suddenly breaking out? Why was it continuing into the winter? Why was it more lethal for subtidal stars? Was the agent killing other animals? How was it transmitted—was it conveyed in their food? How many species of star were affected? Another question with personal significance rapidly emerged: were the starfish in the San Juan Islands sick as well?

That night at Alki Beach on December 17, 2013, is embedded in my mind as the moment when the real craziness began for us, when we realized that starfish mortality was spreading rapidly and cropping up in unpredictable places, and we began to worry about our local populations of starfish. That same week, Laura James and Ben Miner (of Western Washington University) saw many ochre, mottled, and sunflower stars dying in deeper water on a dive near Mukilteo, Washington. Deaths were reported in lower Puget Sound; increased mortality was reported by Steve Fradkin from Olympic National Park; and reports kept coming in from Melissa Miner and Pete Raimondi's monitoring program in California and the Vancouver Aquarium's network. Shaken by the carnage at Alki, news of these new mortalities, and the observations from Monterey of the wide species range being affected, I returned to my home in Friday Harbor in December worried that starfish in the San Juan Islands would also sicken.

Morgan Eisenlord, my assistant, was still in Friday Harbor, and she was as determined as I to crack the mystery of what was killing the stars. A week after our work at Alki Beach, on that same low tide series in December, she and I went north to check the status of the stars in Friday Harbor. The low tide was after 10:00 pm, but it was a relatively warm, still winter night, about forty degrees with no wind or rain. We started with an easy place as the tide was dropping. We walked down the pier at the lab docks and shone our lights on the pilings, a little nervous about what we might see. It was low tide, so the pilings were uncovered and we could immediately see about six sunflower stars, each one almost as large as a manhole cover, lurking on pilings just at the water line. Then we headed back up to the car and decided to drive over to check the town docks. In relief, we counted another fifteen sunflower stars on the town docks, each of their roughly twenty arms all looking straight and not twisted.

We drove out of town to a more remote intertidal shore on the northeast side of San Juan Island known to have high populations of ochre stars. We clambered down toward the shoreline over huge boulders in the dark, moving slowly over treacherous, slippery, algae-covered rocks with just the light of our head-lamps to guide us. This site had no beach and was a jumble of huge, car-sized boulders. After an hour of surveying three sections of the shore, we counted forty-seven healthy ochre stars, fifteen small healthy sunflower stars, and a few healthy rose stars. Best of all, we saw no infected individuals. Phil Green, the Nature Conservancy preserve steward on remote Yellow Island, just across the water, also went out that night and reported all was well with the stars there. The next night, Morgan took a friend and surveyed another site, one on Orcas Island with thou-sands of stars. They were all fine. We were relieved but baffled.

Why were the San Juan Island stars still healthy? Could it be that they were immune? Our ochre star populations had experienced a similar wasting disease event in 2011 and we lost quite a few stars that summer. What if the survivors from that event were now immune and had produced the hardy population of today, somehow resistant to the disease? There were other possible explanations, but we liked the hypothesis that our stars were somehow resistant, since it suggested an evolutionary process of natural selection for resistance to disease.

We had reached the end of December, and my scientific colleagues and I still didn't know what microorganism was causing the starfish disease outbreak. We didn't even know for sure if the starfish disease signs and deaths were caused by an infectious agent. We didn't know how it was spreading between stars. We had no idea where the mystery microorganism—if there was one—was emerging from. While pondering all these questions, we were concerned that stars were continuing to die at sites all around us, and we wanted to know why.

The public was concerned as well. People were seeing dead and dying stars at the shoreline and the die-off was getting attention from the media. In early January 2014, Katie Campbell, the PBS reporter who had called me at the conference in California and then came to interview us in Friday Harbor, developed a three-minute video story introduced by Gwen Ifill on the PBS evening news. That clip would eventually get nearly one million online views and set Katie up to win an Emmy Award for her next report about the starfish outbreak.

· · ·

Even though most of my colleagues and I were convinced that we were dealing with an outbreak of infectious disease, consensus was still lacking in the scientific community at large during the winter of 2014. In February, Bruce Menge, professor at Oregon

State University, told me that all the stars at their sites in Oregon were healthy. At a scientific meeting in February, I told him of the progress Ian was making in identifying a causative microorganism, but he was unconvinced that it was an infectious disease outbreak. He wondered whether stars in at least some places were dying from a strictly environmental cause like low oxygen or ocean acidification. He argued it was not likely that an infectious agent would cause big outbreaks in some places and leave intervening populations, like those in Oregon and San Juan Island, healthy. He had a good point: the facts were confusing. I was hard pressed to explain how an infectious agent could simultaneously start killing stars on Washington's outer coast, sites north of Vancouver, and in Monterey, then spread to Seattle, Mukilteo, and Whidbey Island, and yet leave our San Juan Island stars and his Oregon stars untouched. Meanwhile, star mortality was spreading very fast south of Monterey. At the same meeting, I heard new reports of a catastrophic and rapidly spreading starfish death wave south of Santa Barbara.

In the absence of anything definitive, many different theories formed about how and why the outbreak was occurring. Some suspected that the cause of the starfish mortality was the Fukushima nuclear disaster that followed the catastrophic tsunami in Japan in March 2011. I received countless emails from nonscientists stating authoritatively that Fukushima radiation was the cause. They wanted to know why we weren't looking into this issue. Some accused us of actively covering up information about radiation pollution causing mutations in animals on our shores. We did investigate the possibility that the Fukushima event was somehow causative, and many other scientists did think hard about this possibility, but there was never a credible linkage. Kim Martini, an oceanographer at the University of

Washington, and various federal labs carefully monitored United States coastal waters after Fukushima and reported no significant radiation. They also regularly monitored local fish and detected no signature of radiation in the fall of 2013 or before. We concluded from this that neither the stars nor the microorganisms in our waters could have been irradiated to a level that could trigger a widespread die-off or make stars susceptible to an infectious disease epidemic. However, there was a big outfall of docks and other wreckage from Japan's tsunami that came ashore in Washington, Oregon, and California in June of 2012, so one possible linkage was that an infected star or some kind of microorganism was conveyed in the wreckage to our shores. We rejected this hypothesis, however, because there were no epidemics of sick stars in Japan in 2012 or 2013, so there was no indication it started there. Moreover, there was evidence that the epidemic had begun in 2011 on the East Coast of the United States.

We very seriously explored the possibility that something in the environment was either the direct cause of the die-off or was somehow triggering an infectious agent. There was nothing obvious. The beginning of the starfish event in fall 2013 did not correspond with unusually warm or unusually cold water, or cold upwelling, or low oxygen. In the Northwest, a lot of stars died first near urban areas like Vancouver and Seattle. But some stars also died in more pristine locations such as on the outer coast of Vancouver Island, in Olympic National Park, and in central California. There were hints at some sites in Washington state—like Alki Beach and Mukilteo—that more stars were dying near polluted freshwater inputs. Fresh water could indeed have direct adverse effects, since starfish have no kidneys and are thus unable to regulate their inner salinity levels. Furthermore, rivers and streams can carry land-based pollutants into the ocean. At the time, several researchers hypothesized that freshwater inputs could facilitate disease either by stressing starfish and rendering them immune deficient or by conveying infectious agents and pollutants from land. This hypothesis had possible ramifications for the site at Alki Beach, which is near where the Duwamish River empties into Puget Sound. The Duwamish contains refuse from an EPA Superfund site and is still considered a severely polluted waterway. However, despite the historic and ongoing pollution from the Duwamish River, Cove 1 at Alki Beach was a teeming star-filled galaxy before the outbreak. Laura James made videos at the site in November 2013, just weeks before the mortality event, showing sunflower stars of all sizes marauding across the sand, eating clams, and scavenging anything they could find. Morning sun stars chased the occasional rose star, along with mottled stars, brittle stars, vermillion stars, blood stars, and striped sun stars. Ochre stars, mottled stars, and sunflower stars climbed the wooden pilings and cleaned off the barnacles and mussels and baby sea urchins. So, despite inputs of pollutants, the site was historically resilient, making freshwater input an unlikely culprit.

While we considered all the competing hypotheses and the evidence that might support or refute each one, our experts on the East Coast were continuing their investigations into possible causative pathogens. At the Bronx Zoo, Harley Newton had pored over hundreds, maybe thousands, of slides from our sick stars and found nothing: no protozoans, fungi, or rickettsial bacteria that would show up in light microscope sections. They were not there. In Ithaca, Ian Hewson and his students had finished sequencing DNA from hundreds of samples from the West Coast. They had identified a virus that Ian suspected was making the stars sick. It was present in most of the sick stars and only some of the healthy-looking ones. This newly identified virus was in the family Parvoviridae. Called a densovirus, it was a very plausible pathogen because it was similar in genetic structure to a known killer: parvovirus of dogs.

Parvovirus of dogs is highly contagious and causes what is also considered a wasting disease. It kills in two ways, by either attacking the digestive system and causing vomiting, diarrhea, weight loss, and anorexia, or by attacking the heart muscles and causing cardiac arrest. It has caused lethal, rapidly spreading epidemics, particularly in puppies. Parvovirus of dogs is transmitted either by direct contact with an infected dog, or indirectly via a fecal-oral route. Heavy concentrations of the virus are found in an infected dog's stool, so when a healthy puppy sniffs an infected dog's stool, it may contract the disease. The grim aspect of its biology is that it has a huge environmental reservoir; there is evidence that the virus can live in soil for up to a year. A massive deadly outbreak in dogs started in 1978 and spread worldwide in one to two years. Stricken pet puppies, strays, and hugely valuable show dogs would develop horrible bloody diarrhea, lose even the lining of their intestines, and die rapidly. In a brilliant technological feat, the parvovirus was identified and a vaccine that now protects most puppies and dogs was developed by Cornell veterinarians.

In January 2014, Ian had accumulated more evidence that the densovirus might be the killer, but a challenge infection experiment was needed to support the hypothesis. Ideally, such an experiment would be designed to fulfill Koch's postulates, the gold standard for establishing the causative agent in a particular disease. It was the procedure Garriet Smith and colleagues had followed in showing that *Aspergillus* was the agent that made sea fans sick. But there was a problem. Viruses live inside cells, and the only way to create a pure culture of them is to grow them in special cultures of specific host cells. Since there are no echinoderm cell lines and it's a big feat to develop them, Ian could not isolate and culture this densovirus. This meant that we couldn't fulfill all four of Koch's postulates. But we felt the work was close to a breakthrough and that it would be convincing to at least verify two parts of the story: that the disease was transmissible between stars, and that it was caused by a virus-sized agent. In essence, we would inject healthy stars with the virus and then see if they got sick.

We had the expertise in my lab to do this experiment. Dr. Colleen Burge was a postdoc working on our coral health project; she had previously studied oyster herpes virus disease and knew how to handle viruses. She had done her doctoral training with Carolyn Friedman, who had identified the causative agent of abalone wasting syndrome. But where could we do the experiments? We needed a big lab space with lots of running seawater, but we were certainly not going to do the work at Friday Harbor Labs. We had no large-scale quarantine facilities to run experiments with a killing disease in a region where all the local stars were healthy. But not far from Friday Harbor, at Marrowstone Point near Port Townsend, the U.S. Geological Survey has a well-quarantined fish virus lab. I approached the lab's director, Paul Hershberger, one of the members of our Research Coordination Network for Ocean Health and a fish pathologist who works on commercially valuable species like salmon and herring, and briefed him on our situation. He graciously offered us his quarantine facilities to run the critical experiment on our spineless starfish.

Colleen agreed on rather short notice to take up residence at Paul's lab and try the challenge experiments. Morgan was also excited by the opportunity to go along and help, seeing it as a big opportunity to learn new lab techniques from an expert like Colleen for handling infectious agents. The Marrowstone Point lab is composed of four buildings stuck out on a sand spit in upper Puget Sound. It is a stunningly beautiful place: you can look across the sound and see snow-capped Mount Baker and Mount Rainier looming on the horizon while bald eagles circle overhead. It is also an isolated place, and use of the internet is very restricted. After spending time there, some of my students joked that all there was to do when not working was watch the eagles and make dolls out of kelp.

The facilities at Marrowstone Point are excellent—a huge volume of running seawater, large aquaria, a whole room for control animals and another whole room for experimental ones. It is designed as a fish pathology lab, so every detail of quarantine was considered. The influent and effluent are treated, so we didn't need to be concerned that we were either drawing in or sending out contaminated water.

One obstacle in running these experiments was that our research team needed native stars that had not been previously exposed to the virus. We had decided to work on the sunflower star, since it was clearly the most susceptible. It had been the first to die north of Vancouver in August 2013 and in the multi-species mortality in Monterey in early December. It was beginning to die through the winter in some locations near Puget Sound, so we felt very pressured to get experiments under way before all the stars in our area became infected. Colleen and her father collected our original healthy stars from Hood Canal in the middle of the night in January, just two weeks before the same site experienced an outbreak. Ben Miner and other divers collected more sunflower stars for us from nearby Whidbey Island, and we collected the rest from the San Juan Islands.

The sunflower star can have twenty-six arms and is fast and hungry. An adult can be up to three feet across, can eat twenty clams per day, and moves at a rate of six inches per second. It is both the biggest and fastest star in the world. For an animal that moves using hundreds of tiny hydrostatic tube feet, six inches per second is impressive. Sunflower stars are also quick learners. Captive ones know when it is feeding time and climb up the edges of their aquaria in anticipation. They are also agile and coordinated—a sunflower star can catch a clam tossed into its aquarium like a dog catches a ball. Underwater in the wild, the really big ones—almost the size of Labrador retrievers—can look a little scary as they silently glide across the bottom or along some deep rock wall, seeking prey.

In March 2014, Colleen had filled all the available tanks with healthy sunflower stars. Consulting with Paul Hershberger, Carolyn Friedman, and Ian Hewson, she made plans to test whether Ian's candidate virus was actually causing disease. The simplest start was to test the densovirus fractions that Ian had prepared and sent frozen. Colleen injected those fractions into the stars and waited over three weeks to see nothing happen. Colleen then developed a protocol to isolate a virus-sized fraction directly from sick stars. Following Ian's guidance on tissue specificity, Colleen ground up tube feet and epidermis, first with a mortar and pestle and then with a Stomacher lab blender, and added water from a tank of infected stars. She then successively filtered the mixture until she had isolated the fraction that was virus-sized: everything smaller than 0.22 microns. If stars injected with this virus-sized fraction developed signs of disease, this would indicate the likely involvement of a virus. Then she made a control injection by carefully boiling the exact same filtrate in its tube, getting it hot enough to kill any active virus. Colleen carefully injected five stars with the live viral fraction and five stars with the heat-killed control fraction, then put each in its own separate aquarium.

Weeks had gone by setting up the experiment, and Colleen was about to start a new postdoc position, so her time was running out. Fortunately, Morgan Eisenlord could come over from Friday Harbor to help and then take over the monitoring. Day by day Morgan carefully fed and checked each of the five stars injected with the virus and each of the five control stars. Each day she called us and reported. Each day nothing much happened. Then she noticed that arms on some of the experimental stars had started to twist. Then, on day fourteen, lesions appeared on two of the stars that had been injected with the live virus inoculum, while the controls remained healthy and untwisted. The next day, two more of the experimental ones were sick. Finally, all the experimental stars had twisted arms and lesions, and all the controls were still healthy. Despite being sorry that our stars had died, we were relieved by the outcome. It had taken several months from planning in January until a successful experiment in late March, but the result seemed clear. We had the first solid experimental evidence of what was killing millions of stars. In our experimental tanks, it looked like a live virus was the causative agent and was transmissible by injection.

The problem was that scientifically, we were walking on thin ice. To an ecologist, replication is everything. Colleen and Morgan had done the experiment once successfully, but with only five animals per treatment. We were unusually lucky in that the results were consistent—all the experimentals got sick, and all the controls looked healthy—but it was a very low sample size, too weak to publish. We had to repeat the experiment if we wanted to publish it and thereby report that the West Coast starfish die-off was caused by a virus infection. Discussing it over on the phone, both Colleen and Morgan agreed it was a good idea, but we all knew it was risky. We were having a tough time finding healthy animals, and what if we re-ran the experiment and it didn't work? It could wipe out our confidence in the first experiment. There were many reasons it might not work again. Our study animals might all become naturally infected, the virus might be very finicky, we might change some crucial detail, or some aspect of the environment might not be right the second time around. Despite all the uncertainty, I felt we had to do it. We had to do the best possible science, even though Morgan might go stir crazy at Marrowstone Point and start crafting kelp dolls. We conferred with Ian and he suggested we try it as what microbiologists call a passage experiment—use virus from the five stars sickened in the first round to create the inoculum to inject for round two.

This experiment, we realized, could have an important secondary purpose. We could use samples from this experiment for a class project for the course we were running for doctoral students at Friday Harbor Labs the next summer. Steven Roberts, who was in charge of teaching the part of the course on using genomics to study the immune system, wanted us to extract immune cells from both sets of stars that we could later use to determine if the immune system was responding differently in the virus-injected animals. This would not only provide the first experimental evidence about how an immune system responds to a real pathogen, but also additional confirmation that the starfish immune system recognized this virus as a threat. We decided to do it.

By now Colleen was working full time in Carolyn's lab in Seattle, and she commuted over to set up the experiment with Morgan. We still had healthy stars in the San Juans, so I sent divers to collect live, asymptomatic animals from San Juan

Island, and I drove them to Marrowstone in big coolers. Morgan and Colleen ran the entire thing all over again, this time with ten control stars inoculated with a heat-killed fraction and ten inoculated with the live virus fraction. Morgan stayed to monitor and care for the stars. This was really a nail-biter; every day we anxiously awaited Morgan's report. Once again, day after day nothing changed. All stars were asymptomatic and happily eating their clams. During this waiting period, Colleen rapidly developed a protocol for extracting coelomic fluid from the stars and spinning it down in a centrifuge to isolate the coelomic immune cells that we would use in the class project. We did not want to influence the outcome of our big experiment by drawing coelomic fluid from the stars, so decided to wait until a star was unambiguously sick to do this sampling.

On day nine Morgan called and said arms were twisting on some of the stars in the experimental tanks and the controls were still normal. Maybe it was working. We waited anxiously over the next days to hear more, but nothing changed. Finally, on day fourteen Morgan called to report more twisting arms and two experimental stars that were showing lesions. Then it all happened at once: the other eight all developed lesions over the next few days. I was nervous that the controls would show signs of disease as well, but they all held steady for the course of the experiment. By day sixteen in early April, it was clear we had a successful run of experiment two. We were now confident that a viral-sized fraction was causing the wasting disease in sunflower stars. We could not say from this experiment which virus it was; it still remained to be confirmed that Ian's densovirus was the specific culprit. But the experiment did show an increase in levels of the densovirus in the injected stars, so all signs pointed to it as the causative virus. To definitively demonstrate that the densovirus was the killer was going to take some very clever science that was, at the time, beyond us; with no echinoderm cell lines, there was no way to cultivate that single virus in pure form.

Still, this was a huge breakthrough. Three months of anxiety about whether we could do the science we needed and rush to stay ahead of the outbreak was replaced by extreme satisfaction that Colleen and Morgan had gotten the essential job done. We had solved a vital part of the mystery, even if many questions remained. Ian's superb molecular virology had shown a candidate virus and Colleen had proved that a transmissible virus-sized fraction was infectious.*

. . .

Identifying the likely starfish killer was a major accomplishment, but it didn't provide a way for us to stop the progress of the epidemic. Following the successful experiments at Marrowstone Point, we learned more about the factors that seemed to mediate the virulence and impact of the virus, but for the most part we had to sit back and watch the outbreak continue its deadly march.

Back in Friday Harbor, on an unseasonably warm April afternoon, the maintenance guys accidently turned off a valve to an outside tank that housed a few sunflower stars. With no input of fresh, cool water, the temperature spiked in the tank before we

---

*The team, with Ian as lead author and a whole lot of co-authors who provided samples or did lab work or experiments, published a paper in an important journal (*Proceedings of the National Academy of Sciences*) documenting the virus as the causative agent (Hewson et al. 2014). It was a real tour de force on Ian's part to do all that research and publish just a year after the actual outbreak was discovered.

noticed. A few days later, arms on the stars started to twist, and a few days after that, lesions appeared. All the stars in the tank proceeded to die gruesomely, the first confirmed cases in our waters. Starfish are not as warm and cuddly as dogs, but they are interactive and interesting to watch, and each individual in the tank had unique behaviors. They had become like pets for me and it hurt a lot to watch them die.

The trauma notwithstanding, this accident provided some critical information. We surmised that when we had collected the stars in the cold waters around San Juan Island they were already infected, and the warming event in the tank activated the disease, either by stressing the animals or by creating a better environment for the virus. This was the first indication that warm temperature might facilitate the disease, and it showed that the virus was now present in the area of the San Juans. A month later, as waters warmed above 12°C, more cases started to appear in nature. Divers saw sunflower stars with lesions under our docks and, in late May, the ochre stars finally began to fall off the rocks in the San Juan intertidal. Since Morgan—who had now been accepted as a PhD student in Cornell's program—had continued to monitor the ochre stars at ten sites throughout the winter, we had excellent baseline data from before the epidemic hit.

That summer, at every site we monitored around the San Juan Islands, the dominant stars in our waters, ochre and sunflower, died rapidly. Nowhere was the die-off more striking than off Eastsound on Orcas Island. For reasons that none of us understand to this day, the deep fjord and sheltered bay off Eastsound were home to the motherlode of ochre stars before the epidemic. The bay is right next to the one main street in downtown Eastsound, a sleepy island town that comes alive with tourists in the summer. I first visited the site on a sunny, hot low-tide day in June 2014, before any stars showed disease signs. At low tide, the gravel and cobble beach stretched away to a tiny rocky island. As Morgan and I walked across the beach, we began to see stars scattered on the beach, many with tube feet clutching clams. The beach was loaded with clams and I remember thinking that the plentiful food supply might explain the abundance of stars. Under rock ledges and crammed into crevices, there were aggregations of twenty, thirty, and forty stars each. We counted over 300 stars on our three transects that day and estimated there might have been 5,000 stars in and around the site. The colors of stars that day were unlike those anywhere else I have seen in the San Juans—an amazing rainbow of the usual purple, joined by orange, pink, and even brown. When I picked up adult stars, I found ochre star babies beneath, sometimes as many as ten per adult. Each one was brownish and the size of a dime. It was a starfish nursery area. What was strange was that the babies were not the offspring of the adults under which they were hiding. Ochre star larvae have a mandatory period of at least three weeks during which they float with the plankton, living it up eating phytoplankton and washing around with the tides. Only after this larval stage do they settle to the bottom and metamorphose into juveniles. So the babies I was finding may have journeyed tens or hundreds of miles to come there. Not long after this visit, this site was heavily hit by the epidemic. By the end of the summer, Morgan was finding very few healthy stars in her surveys. It was easier to look at the graph of the decline than it was to record the sad week-by-week toll the disease took on the population at this remarkable site.

Our data showed a catastrophic decline of both the intertidal ochre star and the subtidal sunflower star around the San Juans in 2014. The declines we observed were mirrored by those detected by citizen-scientist divers in the larger trans-boundary region reaching from Puget Sound north to Vancouver Island. Ten years before the outbreak, a dedicated cadre of divers who love exploring these cold and challenging high-current waters had begun collecting consistent data on the most common species seen in those waters, and they continued their surveys through the outbreak of 2014. Their data tell the full story of the catastrophe because they capture the historic population levels and comprise the whole regional context. That the full value of their data was realized only after an unexpected die-off demonstrates the importance of regular monitoring of populations and environmental conditions.

The summer of 2014 yielded grim news in other locales beyond Washington State. I received word that summer from Bruce Menge that all the stars at his sites in Oregon were dying rapidly. He was now willing to consider infectious disease as a cause and would go on to lead a paper in 2016 about the impact of the disease epidemic in Oregon.

By year's end in 2014, the outbreak felt like a message from the nature overlords: ignore infectious disease in the ocean at your peril. Most of our previous outbreaks hit low on the food chain and in deeper water. None of us could have imagined a lightning strike hammering the entire guild of West Coast starfish, but that's what was happening. As an invertebrate biologist who prizes biodiversity, I found it nearly impossible to bear the loss of so many beautiful, iconic starfish species all at once. I was somewhat inured to the "death by a thousand cuts" scenario of gradual loss of ocean biodiversity, but a loss so huge, horrific, and ecologically valuable in one event hit home in a devastating way.

.　　.　　.

Since 2014, Morgan and our student interns have worked at a furious pace to survey starfish populations at field sites, deploy temperature loggers, and run key lab experiments to figure out what environmental factors might be making the outbreak worse. Our lab experiments and field observations both point to a clear pattern: warmer conditions have made stars sicker and caused them to die faster.

The summer of 2014, fall of 2014, and entire year of 2015 were the warmest on record at the time. In 2015 a blob of warm water sat in the northern Pacific for months and brought anomalously warm water to our shores. That winter, people reported to me that Alaska still had flowers blooming in January. The outbreak as mapped by Pete Raimondi and Melissa Miner's MARINe group extended from Mexico to Alaska (see map 6). The record-breaking years of 2014 and 2015 preceded the even hotter El Niño year of 2016, which was devastating around the globe, especially for coral reefs. Throughout these years, we think it was temperature that shaped the larger patterns of the outbreak, causing extirpation of the ochre and sunflower star from the warmer waters of California and their survival further north in the colder waters of Alaska. Our lab experiments, described in a paper that Morgan published in 2016, clearly showed that stars died faster in warmer water. The importance of temperature was later confirmed in several other studies, including our own work with sunflower star mortality. The continued warming of the Pacific and of the whole planet does not bode well for starfish.

And it is not just starfish that have been affected. As a result of the loss of the ochre and sunflower stars, both keystone predators, the intertidal and subtidal zones along much of the West

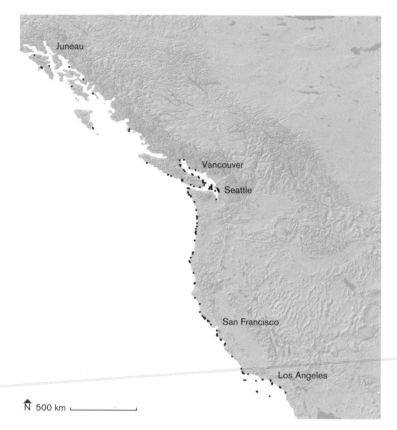

Map 6. Starfish epidemic locations (compiled by Seastarwasting.org).

Coast of North America are now different places in terms of their ecology. In addition, in ways that are harder to quantify, these near-shore environments have lost something of their soul, their gestalt. We hear frequently from kayakers who miss the rainbow of intertidal stars, intertidal beachgoers who see fewer stars on their beaches but send memories and photos of the years before the die-off, and scuba divers who are still looking for the return of the sunflower star.

The almost complete loss of the mighty sunflower stars and several other species in deeper waters has allowed a huge influx of sea urchins, whose populations from California to Alaska had previously been controlled by the voracious sunflowers. The hordes of hungry sea urchins have decimated kelp beds in California, Washington, and some locations in British Columbia, turning them into pink algal pavements called urchin barrens. They are very much like over-grazed cattle pastures (see figure 11). This effect is a trophic cascade similar to what Bob Paine observed with ochre stars and mussels in his experiments on Tatoosh Island: a top keystone predator is removed and populations of the lower level herbivores once controlled by the predator explode and mow down the producer organisms at the base of the food chain. The alarm about the decline of the kelp beds is widespread on our coast. In California, the news media has reported on the situation with headlines like "Scientists and Fishermen Scramble to Save Northern California's Kelp Forests" and "Collapse of Kelp Forest Imperils North Coast Ecosystem." Tristan McHugh's talk in November 2018 at the Western Society of Naturalists meeting showed the data linking sunflower population explosion and kelp decline. The once-extensive kelp forests are major nursery sites and habitats for valuable fish species and many important components of our marine ecosystems. Will the sunflower stars be able to return and roll back this change?

· · ·

What about the future? We still have big questions about the starfish outbreak that make predictions difficult. We don't know the extent to which the virus is still involved in the continued die-off, and what role environmental factors are playing as facilitators.

Figure 11. Urchin barrens on Sakinaw Rock in Sechelt Inlet, British Columbia. After the loss of sunflower stars, green sea urchins reproduced unchecked, transforming a rich habitat (January 2017, upper photo) into an area dominated by hungry urchins (October 2017, lower photo). Photos by Neil McDaniel.

Observations about starfish population trends are mixed depending on locale and species. The autumn of 2017 saw an upsurge of starfish deaths in certain locales. On the other hand, the ochre stars in our waters around the San Juan Islands are now rebounding, and disease levels are very low. The counts of ochre stars at several of our sites, including Eastsound on Orcas Island, were increasing in the summer of 2017. The new appearance of baby ochre stars in southern California gave some hope even for warmer parts of the range. A paper published in 2018 by a team of scientists from the University of California at Merced, led by Laura Schiebelhut, reported some very good news: big increases in populations of the ochre stars and newly recruited baby stars that were growing up. Most exciting, their complete genetic analysis of stars before the outbreak compared to after the outbreak showed a big genetic change. They suggested that the epidemic had killed all the susceptible stars and the survivors were hardy genetic stock that was resistant to the virus. The bad news is that new surveys in 2017 show that the sunflower star is virtually gone from California to Alaska. Our 2018 analysis shows that the magnitude of the ongoing decline from California to British Columbia is linked with the ocean heat wave that hit our waters between 2013 and 2016 (see figure 12). New data from NOAA show a complete absence of the sunflower star from waters as deep as 1,100 meters, so there is no deep water refuge (Harvell et al. forthcoming).

What explains the different outcomes for these two star species, that the ochre survives and increases and the sunflower continues to decline? We think it is the expected action of a multi-host pathogen. The sunflower star was the most susceptible and for some unknown reason does not have the genes for resistance; or perhaps it's that the sunflowers with those genes are very rare and it will take them a lot longer to come back, if they come back at all. We do not currently have answers to these rather urgent questions about the evolution of disease resistance in the ocean.

Whatever the eventual outcome, the starfish epidemic has been a turning point for me and many others. After watching ocean biodiversity decline for most of my career, the near total loss of some

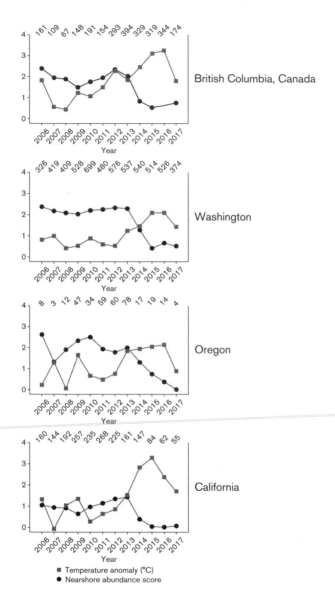

Figure 12. Decline of the sunflower star during a marine heat wave. Compilation of REEF Citizen Science data (Janna Nichols and Christy Semmens) and temperature anomaly data. Numbers across the top are the numbers of transects surveyed. Figure by Diego Monticeno, Jamie Caldwell, and Scott Heron (adapted from Harvell et al. forthcoming).

of our most iconic and ecologically important invertebrate species in such a short span of time was a blow. With the threat of the imminent extinction of tropical corals, white abalone, and our sunflower star, and their cascading ecological impacts, we crossed a line in the sand. As a society, we can no longer stand by idly. We have let loose many destructive forces over which we have little control, but science can still make a difference. The problem is that inadequate funding remains a major impediment to scientists identifying the causes of disease outbreaks in the ocean, investigating the contributing factors, and developing solutions.

To research the starfish epidemic and identify the causative agent, we needed millions of dollars. I couldn't imagine how to shake loose that funding on that scale for immediate use, so we went at it with only small grants from the National Science Foundation and Cornell. It was telling that one source of funding that proved important was a fundraiser called Save Our Stars that was conceived of and carried out by Mrs. Bailey's sixth grade class at Carl Stuart Middle School in Conway, Arkansas. They were so worried about the fate of the stars after seeing the videos that Laura James of Seattle produced that they launched a campaign to raise money by selling t-shirts and small cards that said "Save Our Stars." I received a $400 check from them, and Chuck and I decided to match their donation, after which a private donor tripled the match. With $3,000 and a lot of volunteers, we were able to fund our field surveys. The whole thing would have played out differently if we had had the right resources from the start—not only funds from school children who stepped up to help, but also government funding.

If the starfish outbreak was instead a deadly virus epidemic among humans, like Ebola, we would have had massive resources at our disposal to investigate the pressing questions surrounding the outbreak. Instead, many people acted as if this huge epidemic affecting a keystone species in the ocean was simply a curiosity and assumed scientists would figure it out. But scientists cannot figure out the causes of disease epidemics without funding and investment in needed infrastructure. Disease outbreaks caused by infectious agents are tough nuts to crack and demand our most innovative, expensive technology—whether they affect humans or the biota that surrounds us. Our country is quite good at monitoring diseases in hospitals and diseases of our food crops: the Centers for Disease Control adeptly identifies and types new human diseases, and the USDA monitors terrestrial food safety and outbreaks of new agricultural diseases. At their disposal are expensive instruments for reading the genetic code of microorganisms and the immune status of hosts. Investigating mass mortality in starfish requires the same technological resources—maybe even more, since we have so many knowledge gaps regarding the oceans, and far fewer tools. But funding to adequately monitor new disease outbreaks in the ocean barely exists. The starfish epidemic is just the latest in a series of such outbreaks, and we are terribly underprepared to research those that are sure to come.

CHAPTER SIX

Nature's Services
to the Rescue

The sea, the great unifier, is man's only hope. Now, as
never before, the old phrase has a literal meaning: we
are all in the same boat.

Jacques Yves Cousteau

After we all got sick on Indonesia's Barrang Lompo Island in
2011, we vowed to someday see what was in the water. We got our
chance a year later. As our boat reached Barrang Lompo Island,
I thought it looked like paradise, with white sand beaches, neat,
brightly painted houses, and red, green, and blue fishing boats at
anchor. We could see fish, corals, and sponges through the water
as we passed over the coral reefs and meadows of seagrasses
ringing the island. Smiling children waved, cheered, and called
as they waded from the beach to welcome us. As we drew close
to the town dock, a different picture emerged—plastic bags,
milk cartons, eggshells, and unidentified pieces of trash floated
all around and were piled several feet high underneath the town
dock. It did not smell good. Children waded through the trash to
gather near our boat. We climbed up the ladder to the dock,

147

talked with the well-wishers, and headed toward the marine lab. In contrast to the ocean shores, the narrow dirt streets and yards enclosing the houses were tidy and clear of trash. We heard a sound like an ice-cream truck and turned to see a small, vividly colored open bus careening around a corner. It stopped for us and we climbed on, joining a group of about six people, including three smiling teenage girls with braces on their teeth. "Selamat pagi!" the girls greeted us in Bahasa, and off we zoomed as the girls giggled and said something in Bahasa about how crooked my nose was. It was fun riding the open town bus and we appreciated the welcome. The bus dropped us at the gate of the marine lab, a large stone building that stood across a wide lawn. The last time we were here, we all left very sick, so this time we brought our own lunches. We dropped off our bags and gathered our field gear, scuba tanks, weights, and collecting supplies and headed back to the boat.

Our goal on this trip was to sample the levels of bacteria to see if the water was as polluted with sewage as we suspected. We were still haunted by Bette's case of typhoid and the lingering question of whether the causative bacterial agent, *Salmonella typhi*, was also in that polluted water. Detecting living *S. typhi* in the seawater would make a clear link between threats to human health and threats to coral health. On this trip, I worked with four of Professor Jamaluddin Jompa's students. As I noted in the chapter on coral, Jamaluddin is a brilliant, award-winning scientist at Hasanuddin University in South Sulawesi who trained at James Cook University in Australia. We had worked together for several years on a large marine protected area project, running coral health and fish diversity surveys across Indonesia with his students. Together we had developed a plan for our project here. We wanted to know how many bacteria were in the water, but as ecologists, we were also interested in the question of how those levels changed with environment. The Spermonde Islands were an excellent place to investigate these questions, since there were many separate islands, each with seagrass beds near shore and a ring of coral reef in deeper water beyond the seagrass. We would sample the water on three different islands each at three different distances from shore: at the shore, over the seagrass bed in front of the reef, and at the reef.

On the boat the sampling was straightforward, with no need to even get in the water. We filled pre-sterilized bottles with water and capped them. Within hours, we had collected all our samples and hurried back to Makassar to analyze them in the microbiology lab at the university. Arniati Massinai, a microbiologist, and her assistants poured hundreds of petri dishes, filtered hundreds of the water samples and placed single filter papers to incubate on each petri dish to test for the presence of *Enterococcus* bacteria. The enterococci are a group of closely related species of what we call fecal coliform bacteria, since they live in the intestines of humans and other animals. They are part of the normal intestinal flora of humans and other vertebrates, and some species are pathogenic. Due to the ubiquity of enterococci in sewage, the US EPA uses the presence of this bacteria as an indicator that determines whether ocean waters are safe for people to swim in. Three days and forty-three hours of round-the-clock labor after our sampling, we had our answer.

There were about 2,500 tiny, one-millimeter-sized orange colonies of *Enterococcus* starting to grow on the plates. This meant a score of 2,500 colony forming units (CFU), over seventy times the EPA cut-off of 35 CFU. Although we had expected some pollution, this level was surprisingly high and indicated that the waters were clearly dangerous for human health. We were basically swimming in sewage. Beyond that finding, levels of bacteria were correlated with distance from the shore. The very high level at the shoreline indicated that the contamination source was almost certainly the people on the island. The levels of bacteria dropped off over the seagrass beds and were quite low near the reefs. Was the reduction in enterococci in samples further from the shore a simple result of dilution? Or were there biological processes going on in the seagrass beds that reduced the numbers of bacteria?

Arniati had poured a few petri dishes with a different medium that was selective for the group of bacteria (*Salmonella*) that causes typhoid. There were colonies growing on that medium as well. This indicated the presence of either *S. typhi* or a related bacterium in the water, making it even more dangerous than if only enterococci were present. But we suspected that many other pathogens were in the water as well, and we didn't yet have an exact accounting of them. In the United States, we could have easily sequenced the DNA of all the bacteria present, but we did not have the facilities for that analysis in Sulawesi.

Two years later, Bette Willis, Joleah Lamb, and I surveyed coral health at many sites around Indonesia, from islands near Bali, to both sides of South Sulawesi, to Raja Ampat off the Bird's Head of West Papua. We wanted to take the pulse of health at ground zero in the coral triangle, site of the highest coral biodiversity in the world. We wanted to know not only how healthy the corals were, but to also seek clues to factors might make them sick and that could be mitigated. Joleah, a fearless and tireless diver just finishing up her PhD in Bette's lab, had spent seven years working in small villages reliant upon coral reefs in Southeast Asia. Bette, a distinguished scientist, intrepid diver, and professor at James Cook University, is arguably the world's expert on coral biology. Both she and Joleah were capable of identifying a substantial majority of the approximately 270 species of corals that we encountered. Over the course of several weeks, we spent a lot of time underwater, often doing three dives per day, with each dive lasting well over an hour. At each site we laid down our ten-meter transect lines and focused on the corals within one meter of the line on each side, counting and identifying each coral head and taking note of its health status. I loved the time underwater beside Bette, identifying and learning about all these new, spectacularly diverse coral species. Some sites were as beautiful and interesting as a garden, carpeted with hundreds of species of rare corals in every color and shape, polyps pulsing with life, symbiotic algae working their magic of creating food for the coral from sunlight. These corals carpeting the reef are quiet solar reactors, bursting with the production of new food as their algal symbionts turn sunshine into nutritious carbohydrates. It's hard to believe such beauty and naturally produced solar biology exists on our planet. On sunny days, I sometimes imagine that my own skin is packed with marine algae and able to photosynthesize. But among the spectacular underwater seascapes and beautiful reefs in Bali, Wakatobi, and Raja Ampat, we found many sick corals. Lower levels of coral health occurred in places where sewage levels looked higher and plastic trash was visible on the reefs, including near Barrang Lompo. Joleah and I talked a lot about how discouraging it was to keep studying sick corals and monitoring death and coral decline. What we really wanted to do was focus on ways to make coral healthier. I realized I had spent my career studying ocean life dying from disease and that it was time to develop research solutions and approaches that can improve the health of the oceans that I love so much and that are the life-blood of our planet.

. . .

I have argued in the previous chapters that in an ocean stressed by acidification, warming, over-fishing, and land-based pollution, infectious disease is an under-appreciated threat. What is so uniquely menacing about disease outbreaks in the oceans is that they are facilitated by all the other factors impacting ocean health and can spread like wildfire on their own. While a plastic bottle dumped in the ocean is an affront to beauty, it doesn't multiply once it hits the water like a microorganism does. Disease outbreaks in the ocean are not to be taken lightly because they pose a lethal danger to ocean wildlife, to ocean aquaculture, to humans, and to the functioning of our ecosystems.

What tools and approaches do we have to combat this new, increasing threat? The outbreaks I have described suggest several ways to attack the problem. One obvious approach is to limit the risk of introducing new pathogens into the ocean by controlling land-based pollution and runoff and better treating all sewage. Once a living, multiplying pathogenic microbe finds a home in the ocean and begins to spread, it is darn hard to get it back in the bottle, so it's best to keep it out of the ocean in the first place. Since some disease outbreaks are inevitable no matter what we do on the preventative side, it's also imperative to do what we can to improve our surveillance of ocean health and our ability to detect and respond to outbreaks. I'll have more to say about these approaches later in this chapter.

Joleah had an idea. "For my dissertation on coral health inside and outside marine protected areas, I took data on how much fishing gear of every kind occurred on each transect. I found a lot of plastic gear in the fishing areas, so I also started taking data on how much plastic was on my transects. It looked like the corals wrapped in plastic had higher levels of disease. There is so much plastic in Indonesia; I'd like to take data on the link between coral health and plastics. If we find that plastic is conveying disease and making corals sick, this is a local factor that we can fix." I am kind of ashamed that I did not see that this was an interesting idea. I was so focused on the corals themselves and their infectious syndromes that taking data on plastic seemed uninteresting. I was also so sick of all the plastic in Indonesia that I just wanted to ignore it and work at sites with less plastic. But I wanted to be supportive of this talented young scientist, so I agreed.

So dive after dive, for hundreds of them from Bali to Raja Ampat, we surveyed the live coral and its diseases and the plastics found on Indonesia's reefs. Most people think that all plastic floats until it breaks down into tiny bits, but on the reefs we found chip sacks, rice bags, diapers, and so many single-use plastic bags ensnared in the branches of our corals. Transect by transect, the evidence mounted that certain diseases were much more likely underneath plastic or adjacent to the cuts and abrasions caused by plastic. Skeletal eroding band and black band and white band syndrome are diseases that kill coral, and they were more frequent on our transects when a coral was touched by plastic. According to Jenna Jambeck of the University of Georgia, Indonesia is the second-highest polluter of plastic in the world. In Indonesia, there is a *lot* of coral touched by plastic.

After we wrapped up our work in Indonesia, Bette and Joleah went on to Thailand to help a colleague do coral health surveys there. Then Joleah came to Cornell to do a NatureNet fellowship with me, based on work we were planning in Indonesia and Myanmar. A year later, in collaboration with Doug Rader of the Environmental Defense Fund, Joleah went on to survey the reefs of the Myeik Archipelago in Myanmar, taking data on plastics all the way.

Our paper describing the high levels of coral disease associated with plastic waste was published in *Science* in January 2018. The response was all we could have hoped for in terms of research that could help change the world, or at least help coral reefs to be healthier. Joleah and our colleagues ultimately examined more than 124,000 reef-building corals on 159 transects across 4 countries of Southeast Asia: Indonesia, Myanmar, Thailand, and Australia. Of the corals with trapped plastic, 89% were diseased, compared to 4% when there was no plastic, a twenty-fold increase. Joleah and one of our Cornell undergraduates, Evan Fiorenza, calculated the reef area throughout all of Southeast Asia and projected that there could be approximately eleven billion pieces of plastic entangled on those reefs in the coral triangle.

What is plastic doing to cause these skyrocketing rates? Injury of any kind creates an opening that can allow infectious microorganisms to invade. The living skin of a healthy coral, like the living skin of a human, is very resistant to attack. But puncture or abrasion with a dirty implement is a recipe for infection. Think about us humans: if we step on a rusty nail, we need a tetanus shot. For corals, plastic is a triple whammy: abrasion creates openings in the living skin, dirty plastic carries pathogenic microorganisms that can immediately invade, and some plastics also shade the coral, reducing its capacity to fight back. Our study further showed that the kinds of coral most valuable as shelters for baby fish, the branching corals, are eight times more likely to contract a disease than the smooth massive corals. Every single big news outlet picked up the story, from the *New York Times,* to the BBC, to NPR, to *Newsweek.* Joleah appeared on national television. Some of the creative media titles were "11 Billion Pieces of Plastic Are Spreading Disease across the World's Coral Reefs" (the *Washington Post*); "Billions of Plastic Pieces Litter Coral in Asia and Australia" (the *New York Times*); "A Third of Coral Reefs 'Entangled with Plastic'" (BBC News).*

We were pleased that ocean health was finally getting the attention it needed. While crafting ocean policy is not currently our job, doing the reliable, strong science that creates a foundation for policy change is, and we had hit a home run. Coral reef decline is death by a thousand cuts, and we had clearly revealed the cause of some of that decline and, at the same time, a solution. Our study showed the level of plastic in Australia was 0.4 plastic items per 100 square meters, compared to 25 items per 100 square meters in Indonesia. Australia has strong plastic management and waste disposal on land, so it does not reach the oceans. This shows that instituting waste management on land is a strong preventer of disease in the ocean. We try to stick with the facts based on data we can measure, but there is one idea that needs consideration, even though we have no data for it: if the story is this bad for corals, which are easy to measure since they live attached to the bottom in shallow water and don't move, what about other organisms that come in contact with plastic? Are they also more susceptible to infectious disease? Is the very sad scenario of turtles entangled in plastic and seabirds with plastic –packed stomachs just the tip of the iceberg of harm?

* HRH Prince Charles mentioned our paper in his address for the 2018 International Year of the Reef at Fishmongers' Hall in London, www.youtube. com/watch?v = zDI4YylFlGg (forward to 2:29).

I've focused most of my research on the approaches that exist in the middle ground between prevention and monitoring. Assuming that some pathogens will be introduced into the ocean, and that some of them will develop increased virulence or ways to exploit weakened immune systems, what can be done to deactivate pathogens, to reduce their numbers, and to increase the resilience of ocean ecosystems? It turns out that quite a bit can be done, and much of it involves using nature itself.

In 2014, Joleah joined my lab as a NatureNet postdoctoral fellow to pursue research on reducing the impact of human sewage and pollution on coral health. I've already mentioned one solution, promoting coral reef health by improving plastics waste management on land. Another issue that plagues the coral reefs of Southeast Asia, not to mention many of our domestic reefs in Hawaii and even seagrass beds in Seattle, is human sewage. Taking advantage of the filtration abilities of natural ecosystems seemed a promising way to solve the problem. Plants have been used effectively in many pollution treatment systems because they can deactivate pathogenic microbes in several ways. They can remove the nutrients the microbes depend on, make the chemistry of the water or soil less hospitable to microbes, support resident bacteria that produce natural biocides, and directly kill or inhibit some microbes by producing their own biocides. Seagrass "meadows" have long been known to clean various kinds of pollution from ocean water through at least some of these mechanisms (seagrasses are flowering plants, like most of those on land). In particular, chemicals isolated from seagrass tissues have been shown to inhibit numerous bacterial pathogens that affect humans, fishes, and invertebrates. We wanted to get good numbers on the magnitude of the pathogen pollution problem on the way to measuring the ability of seagrass to reduce harmful bacteria. To accomplish these goals, Joleah needed to repeat more intensively the earlier study we had completed with Jamaluddin and his students and add in more seagrass sampling. She also wanted to combine the pollution survey with a coral health survey to see if seagrass beds improved the health of nearby coral. This project represented a new approach, with relevance for both human health and coral health. If seagrass beds could be shown to be good for both, that would point the way toward solutions. It was in hope of finding that seagrass beds did indeed provide these kinds of benefits—what scientists call ecosystem services—that the World Bank funded part of our research.

Joleah, Arniati, and their team of students from Hasanuddin University departed the harbor of Makassar in a small, somewhat rickety fishing boat and headed back to Barrang Lompo in 2014 with an ambitious plan: to sample along a gradient from shore to reef at four separate islands. At each site they would gather water samples from within seagrass beds and compare them with samples from an adjacent region without seagrass. The hypothesis was that the reduction in fecal coliform bacteria (*Enterococcus*) would be much greater in the seagrass beds than outside. They would also survey coral health to see whether corals inside seagrass beds had fewer diseases than those far away from these beds.

The team tirelessly collected hundreds of replicated water samples from the four islands, which were carefully selected for similar currents and distribution of seagrasses. All the sampling needed to take place in a single day to control for environmental variability; changes in precipitation or wind patterns could affect the bacterial levels detected and render the samples across sites incomparable. Each island took about one hour and twenty minutes to survey: it took the team twenty minutes to travel between islands and twenty minutes to collect samples at each of the six stations on every island. Some sampling at each island was done to try out a plastic dipstick-type "rapid test" for *Salmonella typhi* developed by the USDA to monitor the safety of meat in US factories. With all the samples on ice, they rushed back to the lab at Hasanuddin University in Makassar and immediately worked for another ten hours to filter all the bacterial samples onto the culture dishes.

Despite the long hours of hard work they poured into the project that day, Joleah's drive for strong data got the better of her and she proposed they go back for further sampling the very next day. Joleah has an endearing ability to make even the hardest work fun, and she somehow managed to make a joke of how hard they all worked. The other members of the team started laughingly calling her "bitch slayer." We think this was meant to be an Indonesian variant on "slave driver," but it was never entirely clear. Either way the name stuck. When I look at the quantity and quality of the work they did, I can well imagine the level of exhaustion they endured. In addition to sampling the four islands, they also sampled over the course of twenty-four hours, taking water samples every two hours, and then surveyed Barrang Lompo's coral health inside and outside of the seagrass beds by diving. While they probably did take on too much, this is often the way it is with fieldwork, especially in remote locations. You never know when or how conditions will compromise your project, so you just have to push through while the going is good.

Joleah and her team counted the *Enterococcus* colonies on several thousand petri dishes after their three-day incubation, and the results were so clear that they had an initial answer even before all the data were tallied and analyzed. The water surrounding all four islands was heavily polluted. Some samples gathered outside the seagrass beds exceeded 1,000 CFU, far above the 35 CFU the EPA considers safe. But what about in the seagrass? A clear pattern was consistent across all four sites: there were half as many *Enterococcus* colonies on the samples taken inside seagrass beds as there were on the samples from outside the seagrass (see figure 13). The "bitch slayer" had the numbers to show that the bacteria were dramatically reduced in the seagrass beds (see figure 14).

These results were exciting enough that Joleah returned to Barrang Lompo a few months later with a team of experts from Australia and Monaco. It turned out that the rapid test for *Salmonella typhi* that Joleah and her team had tested was inconvenient to use and ineffective in detecting the presence of *Salmonella*, but they were still not convinced that *S. typhi*, not to mention other very dangerous human pathogens, was absent from the water. Their new goal was to determine definitively what was in the water by sequencing the DNA of every bacterial strain present. The samples that Joleah and her team collected ended up containing about 1,200 different bacterial types or species. We knew the pollution was bad, but we were still shocked at the levels and diversity of human and invertebrate pathogens that were in the water.

Among the classes of bacteria detected in the water and described in our 2017 paper that Joleah led were several associated with human disease, including *Enterococcus* (which cause diarrhea), *Streptococcus* (pneumonia and meningitis), *Clostridium* (tetanus and botulism), *Pseudomonas* (wound infection), *Staphylococcus* (colitis, gangrene boils and blisters), *Vibrio* (diarrhea, cholera, and death), and *Bacillus* (anthrax). Joleah never did find the bacterium we were looking for initially—*Salmonella typhi*—so we concluded that despite its high levels on the islands, *S. typhi* bacteria do not survive well in seawater. This was a bit of good news; there are prob-

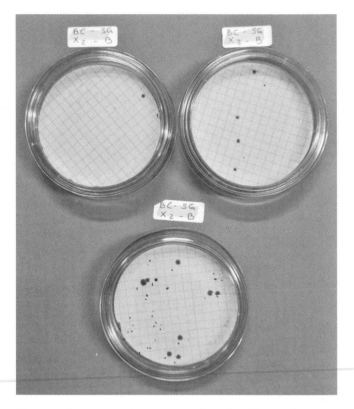

Figure 13. *Enterococcus* colonies on test plates from Barrang Lompo. The three plates on the left are from seagrass meadows; the three on the right are from locations without seagrass. From Lamb et al. 2017.

ably many other harmful bacteria that get filtered out right at the interface of air and salt water. On the whole, though, the genetic-sequence data proved that there is no shortage of very dangerous, disease-causing bacteria in the water, including ones that are known to cause diseases in invertebrates and fish, such as *Shewanella, Pseudomonas, Rickettsia,* and *Pseudoalteromonas.*

Although there were many species of dangerous bacteria present, the fact that their levels were dramatically reduced in

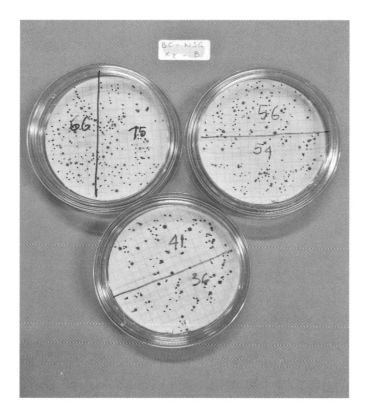

the seagrass meadows was encouraging. Both the levels of *Enterococcus* bacteria counted on the plates and the levels of pathogenic bacteria measured by genetic sequencing showed the same pattern. The seagrass meadows seemed to have a very high capacity to remove pathogens dangerous to humans. If we wanted to quantify the value of nature in benefiting human health, Joleah and her colleagues had done work that could let us put a number on the per-hectare value of a seagrass bed.

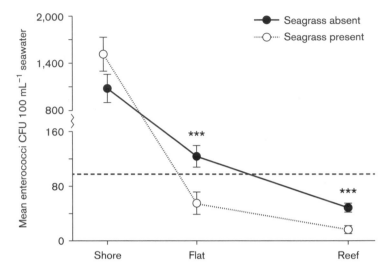

Figure 14. Average counts of *Enterococcus* colonies on test plates from four islands in the Spermonde Archipelago. Water was sampled at the shore, inside seagrass beds (flat), and at the reef, on the seaward side of the seagrass beds. From Lamb et al 2017.

It was also important that the research showed that the corals, the most ecologically valuable foundation member of the natural biota, had significantly less disease in the seagrass beds. In addition to providing further proof of the detoxifying value of seagrass, the findings on coral allowed us to establish that critical link between human and coral health. This gave us a positive and practical way forward. If we conserved and protected seagrass beds, instead of destroying and removing them, we could slow deterioration of ocean health and possibly prevent some human disease, even in remote places like these sewage-polluted islands in Sulawesi. It might even be possible to find ways to *increase* the cover of seagrass in coastal ecosystems and thereby improve the health of both the ocean and nearby human inhabitants.

This study raised many new questions that would be the focus of future studies. How long can the bacteria live in the ocean? How often and at what rates are new bacteria introduced? Through what mechanisms is the seagrass reducing the bacteria? Further, the study results had implications for places outside of Indonesia, because various types of seagrass grow in shallow tropical and temperate ocean waters all over the world. Should we consider manipulating seagrass beds in places like Puget Sound, near Seattle, where there are lush seagrass beds and also annual sewage spills when the rains overrun sewage treatment capacity? Do seagrass beds in Puget Sound work as well as seagrass beds in Indonesia to soak up and detoxify pathogens? How do factors like the density of the seagrass beds, their areal extent, their location, and their proximity to land-based pollution affect their capacity for marine hygiene? Can the restoration of seagrass beds help buffer coastal pollution?

Harnessing plants' natural filtration abilities and antimicrobial properties to reduce pollution is an absolute no brainer and we are not the first to think of it. Both natural and constructed wetlands have long been touted for their ability to soak up and reduce polluting nutrients. More recently, constructed wetlands and riparian buffer zones between agricultural lands and streams have been used to soak up both excess nutrients and toxic agricultural pesticides. Constructed wetlands are also considered valuable and efficient for some kinds of sewage treatment in small rural communities. Our work shows the potential for applying these principles in marine ecosystems. We think that seagrass meadows adjacent to small island communities without adequate septic treatment, like the Spermonde Islands, have a huge potential for mitigating pollution from sewage. We also believe that both natural wetlands and seagrass beds around

urban centers like Seattle and New York City have under-
appreciated value for detoxifying coastal waters when these cit-
ies' quite efficient septic systems periodically overflow during
high-rainfall events.

To optimize the way that we use the natural filtration service
of seagrass beds, we need to fill big gaps in our knowledge. What
we know is that the levels of bacteria passing over a fully functioning, healthy seagrass bed are reduced substantially (about
50%, going by the average of our samples). How does this reduction actually happen? Where do the bacteria go? We believe
there are at least four ways that bacteria are reduced in seagrass
beds: detoxification by oxygenation, detoxification by resident
bacteria (called the microbiome), passive filtration by seagrasses,
and active filtration by resident organisms (clams, ascidians,
mussels, bryozoans, worms). Knowing more about the mechanisms by which bacterial levels are reduced would enable us to
develop optimized artificial seagrass beds that could be deployed
where reduction of pathogen pollution is needed. For example,
if we discover that resident animals like oysters are the main filtering mechanism, we could beef up seagrass beds with battal-
ions of bivalves. Or if seagrass beds work like municipal sewage
treatment systems and kill anaerobic bacteria by releasing oxygen (human gut bacteria thrive in environments that do not have
oxygen), then we could maximize the exposure of pollutants to
seagrasses during daylight hours, when the seagrasses are pho-
tosynthesizing and producing oxygen. In Indonesia, Joleah
measured bacterial levels every two hours and found a build-up
of bacteria during dark hours. This is consistent with the notion
that daytime oxygenation in seagrass beds kills some anaerobic
bacteria. Similarly, septic treatment plants work by using pulses
of oxygenation to kill bacteria. Joleah and I are excited to expand

our research program and continue to study and subsequently rank the importance of these different bacterial fighters in seagrass. We hope to ultimately identify tools that help heal the ocean and not just diagnose disease. We are continuing this work closer to home, to see if the seagrass beds close to Seattle also can remove harmful bacteria.

Another promising ecosystem-services technique is to use bivalves to reduce the pathogen load. Bivalves feed by filtering ocean water through gills designed to collect plankton and edible organic matter. In this way they can remove pathogens as well as food from the water. Indeed, bivalves and other filter feeders that live in seagrass beds may be part of the engine of detoxification in our Indonesia seagrass beds. Bivalves such as oysters and mussels have been used successfully to clean polluting nutrients from harbors and estuaries. In one study, scientists hung strings of mussels from floating rafts in the Bronx River Estuary and found that the "mussel islands" absorbed as many polluting nutrients from the water as commercial recovery efforts.

In an ongoing study, Carolyn Friedman, Colleen Burge, and our collaborator Maya Groner have been investigating if bivalves can filter out the infective zoospores of the slime mold *Labyrinthula zosterae*, which grows within the tissues of eelgrass and creates large black-bordered lesions on the blades. Carolyn called her shellfish farm collaborators and ordered a few thousand baby oysters to do the job. The tiny baby oysters, three months old and five millimeters long, arrived and sat quietly in our sea tables awaiting their experiment, while students grew the eelgrass pathogen in petri dishes. Once the cultures grew, the students planted sterile eelgrass blades in ten sterile beakers and added a few oysters to half the beakers. On the day of the experiment, they introduced a known concentration of pathogen into all the beakers. Five days later they detected fewer lesions on the blades with oysters and more on the blades without oysters. It was a pilot experiment and the data were more variable than they could publish, but it suggested that the oysters do filter out zoospores. Colleen, our starfish virus wizard and now a professor at the Institute of Marine and Environmental Technology in Baltimore, recently reran the experiment with increased replication and confirmed that the oysters do substantially reduce pathogenic cells in the water. This was a small-scale proof-of-concept that bivalves can filter out a pathogen that kills seagrass. The next step is to scale up this experiment and eventually to test it in nature.

Colleen and Carolyn are world experts when it comes to bivalves and the pathogens they can harbor, so they recognize the complexities of using bivalves as biofilters. Filter feeders can either transmit a disease agent or dead-end it. Which of these processes occurs depends on the selectivity of the filter feeder, the degree of infectivity of the pathogen, the mechanism of pathogen transmission, and the ability of the pathogen to resist degradation. Some bacteria and viruses are hardy in seawater and can accumulate active infections within certain filter feeders. In such a case, it is possible to have the perverse outcome in which the filter feeder actually amplifies the disease (this is known as bioaccumulation and is why it is not a good idea to eat raw oysters from polluted water). But there are also examples of filter feeders that effectively clear the water of certain pathogens—like the bacterium that causes bird flu—and deactivate them (see figure 15). Carolyn and Colleen point out that since bivalves like clams and oysters can concentrate microorganisms, they are also useful as sentinels—essentially, canaries in the coal mine—that can be monitored for the presence of pathogenic microorganisms. This is the basis for a national program called Mussel

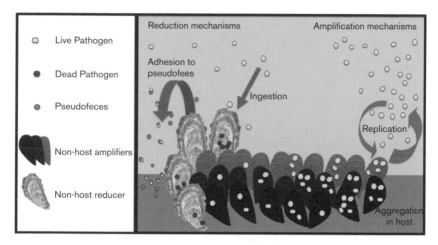

Figure 15. Filter feeders can alter pathogen transmission through reduction or amplification. Reduction can occur via mechanisms such as ingestion by non-target filter feeders (dark arrow) or incorporation into pseudofeces made by the filter feeder that sink out of the water column (curved arrow). Amplification occurs when the pathogen successfully replicates inside or on the filter feeder (two curved arrows at right), or if aggregation in the filter feeder increases the likelihood of transmission. Adapted from Burge et al. 2016.

Watch, run by the Washington Department of Fish and Game, that puts out sentinel mussels as pollution monitors in urban waters. Joleah and I have started a new project with Jennifer Lanksbury (Washington Department of Fish and Game) in Puget Sound to use mussels to monitor bacteria.

. . .

Biofiltration by plants or bivalves is only one example of how humans can harness natural marine ecosystem processes to promote the health of our natural biota, protect human health, and reduce the risk of disease outbreaks. In considering other possibilities, I realized that some of the most important solutions may lie in the kinds of interactions microbes have among themselves.

This is the realm of the microbiome, the hottest new research frontier in both human and ecosystem health. As is so often the case, microbiome science is best developed in the terrestrial world. In his 2016 book *I Contain Multitudes,* Ed Yong explores new discoveries in microbiome biology and their relation to human health. The ideas range from the now widely accepted practice of keeping pets in your home in order to bring in a different microbiome that includes beneficial bacteria, to microbiome balls, little balls seeded with beneficial bacteria that can be transferred to anyone who handles them. Scientists believe that microbiome balls may be a way to seed environments like homes and hospitals with favorable bacteria.

How can we use natural microorganisms to promote ocean health? This is an exciting frontier that has barely begun to be explored. One obvious way to start is to expand the area covered by marine ecosystems like seagrass meadows that are known to house beneficial microbiota, and to study how they interact with pathogenic microbes. Working in temperate eelgrass beds, Yuka Onishi from Hokkaido University discovered a bacterium in the seagrass microbiome that is an active fighter of harmful algal blooms. It lives on the surface of seagrass blades and releases toxins that reduce the growth of the harmful algae. Studies are under way to reveal if having a whole seagrass bed full of these bacteria reduces the risk of a harmful algae bloom.

My favorite example of a new microbiome solution to pathogen pollution and disease is phage therapy. Phages are the viruses that in nature feed on bacteria, so using phages to control bacteria is a form of bio-control, like using spiders or ladybugs to control insect pests of crops. Recall that Carolyn Friedman discovered that the dark purple spots she saw under the microscope when she examined her rickettsia-infected abalone were phages, and that the phages were likely conferring disease resistance to the abalone. Although there are few examples of using phage therapy in the oceans, it has a very long history on land as a cure for human bacterial diseases. During the First World War, before the discovery of antibiotics, an active phage therapy program was developed for humans. People with bacterial infections were intentionally infected with selected viruses to kill the bacteria. In Tbilisi, Georgia, the George Eliava Institute of Bacteriophage, Microbiology and Virology has been active since the 1930s in the field of using phages to combat human microbial infection. The evolution of antibiotic resistance in the United States has brought the efficacy of many of our antibiotics to a screeching halt, and scientists are now looking at phage therapy as an alternative.

Researchers are actively looking into phage therapy as a way of controlling diseases in livestock. In one prominent example, Rodrigo Bicalho, professor of dairy production medicine at Cornell University's veterinary college, is investigating the use of phage therapy for cattle infections for which antibiotic therapy is not efficacious. Bicalho is testing cocktails of phages isolated from farms. To date, his lab has assembled two large collections of bacteriophages: one collection is composed of phages active against the bacterium *Pseudomonas aeruginosa,* which causes respiratory, skin, and urinary tract infections in both humans and cattle, and the second is composed of phages that are active against *E. coli,* which causes gastrointestinal infections. Bicalho's lab is currently conducting a clinical trial of four bacteriophages aimed at preventing skin infections in commercial dairy cattle in New York. Phage therapy could be very useful in livestock production because of the emerging dangers of polluting farms and farm foods with antibiotics that breed resistant bacteria.

Eugene Rosenberg of Tel Aviv University and his students have spent the last decade investigating the possibility of phage therapy for corals. Over several years, they painstakingly isolated thousands of viruses from nature and finally found two strains that actively eat up coral pathogens. They grew hundreds of petri dishes full of these phages and tested them in the lab. Sure enough, levels of the coral disease were reduced when the phages were introduced. The next step was a field trial: they grew even more of these strains of phage and injected them into specially designed field containers containing sick corals. The innovative experiment worked: the corals with phages added lived longer. Although this intervention strategy is promising, it is a long way from being useable as an actual treatment for sick corals in the wild or even in a lab setting. Finding and culturing the right phages is the first step, but introducing them in high enough numbers into the wild is infinitely more difficult. Furthermore, we don't know how long these phages will remain effective, since bacteria can evolve to evade their predators. And, of course, introducing biologicals into the wild may have unintended consequences. Advancement of this research theme requires trained virologists with access to laboratory facilities and commercial infrastructure.

.   .   .

One of the key lessons we've learned from the disease outbreaks I've described in this book is that effective surveillance of ocean health and monitoring for outbreaks must be a part of how we confront the threat of ocean disease. The response to the starfish die-off showed that we are unprepared to handle major disease outbreaks in the ocean. It took us months to even verify that the cause of the die-off was an infectious disease, as opposed to ocean acidification, pollution, or temperature extremes. The slow timing of response set us back in terms of doing the proper sampling to determine what controlled the spread of the disease, and most importantly, to be able to adequately investigate the causative agent. After all, if we can't identify and diagnose the problems, how can we even begin to think about preventing or fixing them?

A framework for health surveillance in our oceans is already in place. The World Organization for Animal Health has established detailed guidelines for handling notifiable diseases and activating an early warning system for outbreaks. Notifiable diseases are those that must be reported to the World Organization for Animal Health when they are detected. For the 117 current notifiable diseases, the organization maintains databases containing up-to-date information about diagnostics and experimental protocols. In the oceans, these diseases include infectious salmon anemia, salmonid alpha virus, abalone herpes virus, and abalone withering foot syndrome. But that's it for ocean diseases. The list of notifiable diseases is dominated by those that plague terrestrial livestock. There is a lot of room to ramp up the ocean end of this global surveillance.

While ocean surveillance struggles to get its due, we invest heavily in surveillance and new technology to get ahead of disease outbreaks on land. In response to widespread recognition that we were too slow to react to both the Zika and Ebola viruses, a new partnership was recently formed to improve surveillance of human disease outbreaks. Donors and drug makers are teaming up to produce a $500 million kitty to further develop vaccines against Lassa fever, the Nipah virus, and Middle East Respiratory Syndrome (MERS). This investment paid off in 2018, when a new outbreak of Ebola virus in Africa was quickly smothered by a combination of early action and a new, successful vaccine. Outbreaks on land *and* in the sea can strike rapidly and unexpectedly. Fast-response capability is well developed for human and land-based agriculture outbreaks but is sorely lacking in the oceans.

To improve our systems for monitoring and responding to infectious disease in the ocean, we might take a cue from what's being done for human diseases. I advocate the use of a report-card-style approach like that developed by the Trust for America's Health and the Robert Wood Johnson Foundation in their 2015 report "Outbreaks: Protecting Americans from Infectious Disease." This report outlines ten key indicators for preventing, detecting, diagnosing, and responding to human disease outbreaks; localities are assessed on each indicator. A similar system could be used to improve protection for our oceans and our coastal waterways. If implemented effectively, this system could allow us to identify our shortcomings and help guide subsequent tracking to monitor improvements.

In addition to putting in place better systems for ocean health monitoring and outbreak surveillance, an absolute priority for confronting the looming threat of marine disease is cutting off the flow of human sewage and agricultural animal waste into our oceans. Many people know the story of the planktonic algae *Pfiesteria piscicida*, the "killer dinoflagellate" that spilled over from hog farms to rivers and estuaries. It turned into both a scourge on natural populations of fish and a human menace that caused debilitating neurological deficits in those who contacted it, from fishermen to scientists. This is a clear example of how runoff from agricultural and aquaculture farms causes dangerous pathogen pollution in our natural ecosystems. It also shows how existing microorganisms can transform into dangerous health threats when the environment changes. The intentional dumping of raw or poorly treated municipal sewage into the oceans may be an even greater offense. I find it particularly galling that the entire municipality of Victoria, Canada dumps all its sewage, untreated, into the richest waters on our continent. The city of Victoria is close enough to be seen from my living room, so it's a snap to imagine those bacteria zooming on fast ocean currents across the waters to my shore. Joleah and I keep threatening to launch a commando raid with our Zodiac across the border to run the same measurements in Victoria's water that we did in Barrang Lompo. Innovative efforts to monitor and regulate pathogen pollution from land will likely start on a state-by-state level, and I predict we will see the most innovative clean water approaches pioneered by California, a state with enormous marine riches and the consistent will to protect them.

I want to see a renewed will to foster health in our oceans. Limiting carbon emissions, doing a better job of managing fisheries and marine protected areas, changing the economic pressures that lead to despoiling of marine environments, and managing population growth will all reduce the risk of infectious disease outbreaks in addition to accomplishing other important goals. To this list I would add an improved focus on developing better science. I have described some of the fascinating new discoveries in microbiology and the powers of natural ecosystems to fight disease that have been placed at our fingertips. It's up to us to get on with working these new discoveries into more resilient management of our oceans.

. . .

It's worth considering what might happen if we don't improve our game. There will certainly be more big outbreaks plaguing our oceans. We don't know yet what will come, but in the afterword I speculate about the possibilities and where we might look for the next big outbreak. What do we expect to cause it and how destructive could it be?

Let's read the writing on the wall and tackle this problem before the trends of biodiversity loss become irreversible, and before we have a truly catastrophic outbreak. Let's put nature's pathogen-fighting services to work by saving and replanting our seagrass beds and setting up biological filtration islands made of bivalves. Let's develop best-practices guidelines and disease-fighting report cards that help us reduce pathogen inputs from land and aquaculture. Let's legislate a fund to handle outbreaks at their very beginnings and not let them run rampant in our oceans. The bottom line for these initiatives is policy change.

We made a good start on this with the drafting of the Emergency Marine Disease Act of 2014. Much to my surprise, the event that triggered the introduction of this new bill in Congress was not a human public health emergency from tainted oysters or clams; it was the starfish die-off. In August of 2014, in the middle of the outbreak, Congressman Dennis Heck's office contacted Pete Raimondi and me for help in writing the proposed legislation. I pulled in Carolyn, and together we outlined legislation that could rally the right expertise and deliver the funding to deal with a new and dangerous outbreak in the ocean. It was introduced in 2014 and reintroduced in 2015. It hasn't passed yet, but it will surface again.

What was it about the starfish die-off that triggered this powerful response? It was an outbreak that people could see on their beaches. It affected an iconic sea creature beloved by children. It happened on both a vast scale and locally. It was a mystery. It was unprecedented. For Mrs. Bailey's sixth grade class in Arkansas, it was terrifying that we could lose a beloved component of ocean biodiversity. For those kids, the oceans are defined by starfish, and to have a star-less ocean struck a deep fear. This response taught me a lesson about the power of public perception. If enough people know what's at stake, if they can imagine what we might lose, they will act. It's my hope that increasing awareness of the ocean's vulnerability and our dependence on it will force a major shift in how we treat this great expanse of our planet. With climate change already impacting agriculture on land and coral reefs in our oceans, and our population continuing to grow, it's urgent that this shift begin soon.

Afterword

The Next Big Outbreak

Infectious disease is nearly as old as life itself. Not long after multi-celled organisms evolved and proliferated in the oceans, some single-celled microbes took advantage of the bodies of these larger organisms as new homes for nourishment, growth, and reproduction. The arms race that ensued between pathogens and would-be hosts spurred the development of ever more elaborate immune systems among the multi-celled life forms and ever more powerful modes and tools of infection among the pathogens. In this age-old battle, victory has always been temporary; over the long term, the outcome has been the dynamic equilibrium of life on earth.

It may be, however, that human beings have inadvertently tipped the environmental scales in favor of the microbes. The outbreaks we've witnessed over the past few decades may only be the beginning of a period of microbial resurgence. Indeed, as my colleague Jeremy Jackson famously said, "We are on the slippery slope to slime." Destructive microbial activity is on the rise. But through scientific research, education, and policy change, we humans may be able to alter some of the forces that are giving microbes the upper hand.

I am looking out across our bay to the city of Victoria, Canada, and considering what might happen if we don't step up our game. What killing force could roll silently into our bay? There will certainly be more big outbreaks plaguing our oceans. Should we be watching for them, and where would we look? What will be the causative agents, and how destructive will they be?

The likeliest future outbreak is one that will strike a farmed species, such as shrimp, oysters, abalone, or salmon. Aquaculture farms can breed pathogens as well as seafood, since they typically contain animals in very high densities. Shrimp, oyster, abalone, and salmon farms are continually bombarded with small outbreaks that have the potential to spread. For example, the Taura virus that affects shrimp has been the world's largest marine pandemic for more than three decades and has fueled massive breeding programs aimed at developing resistant shrimp. There is also an outbreak of oyster herpes virus that threatens the health of oysters on the West Coast and worldwide. Deeper in the ocean, there are lurking salmon viruses that could rapidly cause a serious outbreak and devastate populations of the popular fish.

Although outbreaks of disease in farmed species are economically and socially expensive to manage, dangerous pathogens are also poised to threaten wildlife on a big scale. I worry, for example, about Morbillivirus attacking the endangered resident orca pod that swims just outside my bay. Morbillivirus, a close relative of canine distemper, has repeatedly caused killing outbreaks in seals and can be transmitted to orcas. In 2011, the US five-year status review of southern resident killer whales acknowledged that infectious diseases could impede recovery of this endangered species. Marine mammals, including orcas, are also susceptible to human viruses such as influenza and carry some of the antibiotic-resistant viruses that affect humans. Antibiotic susceptibility testing has revealed multiple antibiotic-resistant gram-positive and gram-negative bacteria in samples from the exhaled breath plumes of killer whales. These bacteria showed increased resistance to multiple antibiotics. Detection of multiple antibiotic-resistant bacteria in this natural setting has significant medical implications for humans who may recreate or work in the ocean, and for those who consume seafood.

While the health of orcas and humans is currently a priority in our waters, the outbreaks that worry me the most are those that suddenly attack foundational marine species and threaten to destabilize the natural ecosystems we rely on. The specter of how rapidly and unexpectedly the starfish outbreak spread still haunts me. One of the other doomsday-like scenarios I can imagine is an outbreak of some kind of pathogen with a very broad host range that affects all crustaceans, those hard-shelled invertebrates often with big claws and stalked eyes. A broad multi-host outbreak could take out all crabs, shrimp, and lobsters, and also the tiny crustaceans called copepods that make up much of the zooplankton. Don't let their size fool you: the copepods in zooplankton are an essential source of food for many types of whales and the entire food chain. They regulate the productivity of our oceans, and when their numbers dwindle, so do populations of the great plankton-feeding whales, like the right whales. An infectious disease of crustaceans with a broad host range would devastate ocean food webs as well as our own favored foods from the ocean.

Even worse than a multi-host pathogen affecting all crustaceans would be an infectious pandemic disease hitting organisms even lower on our marine food webs, like the one-celled plants called coccolithophores. In nutrient-poor waters of the open ocean, coccolithophores are an essential source of nutrition for small fish and all those diverse zooplankton. Coccolithophores are calcified and play an extremely important role in carbon and sulfur cycling in the oceans. *Emiliania huxleyi* is the most abundant coccolithophore globally, and we know quite a bit about its host-pathogen dynamics. Increases and decreases in its population are controlled by viruses. Coccolithophore susceptibility to the viruses is determined by the density of their calcified skeleton and the temperature of the water. So as temperatures warm and acidification continues to increase rapidly in our oceans, will this fundamental plant at the base of our great ocean food chains become more susceptible to its viruses?

At this point in the story, we do not need a crystal ball to see the future. Warming the climate and polluting the sea will give new opportunities to underwater microorganisms, resulting in explosive new outbreaks of infectious disease. The bigger question is, how will we respond? We have transformative technology for disease diagnostics and surveillance on our side that could make a crucial difference if we put it to work. My hope is that we will be fast enough to develop innovative ways to control the eruptions of new outbreaks and save the ocean's biodiversity.

References

1. WHAT RISES WITH THE TIDE?

Alroy, J. 2015. Current extinction rates of reptiles and amphibians. *Proceedings of the National Academy of Sciences* 112(42): 13003–8.

Aronson, R. B., and W. F. Precht. 2001. Evolutionary paleoecology of Caribbean coral reefs. In W. D. Allmon and D. J. Bottjer, eds., *Evolutionary Paleoecology: The Ecological Context of Macroevolutionary Change.* New York: Columbia University Press, 171–233.

Burge, C. A., C. M. Eakin, C. S. Friedman, B. Froelich, P. K. Hershberger, E. E. Hofmann, L. E. Petes, K. C. Prager, E. W. Weil, B. L. Willis, S. E. Ford, C. D. Harvell. 2014. Climate change influences on marine infectious diseases: Implications for management and society. *Annual Review of Marine Sciences* 6: 249–77.

Carpenter, K. E., M. Abrar, G. Aeby, R. B. Aronson, S. Banks, A. Bruckner, A. Chiriboga, et al. 2008. One-third of reef-building corals face elevated extinction risk from climate change and local impacts. *Science* 32: 560–63.

Gire, S. K., A. Goba, K. G. Andersen, R. S. Sealfon, D. J. Park, L. Kanneh, S. Jalloh, et al. 2014. Genomic surveillance elucidates Ebola virus origin and transmission during the 2014 outbreak. *Science* 345: 1369–72.

Harvell, C. D., R. Aronson, N. Baron, J. Connell, A. Dobson, S. Ellner, L. Gerber, et al. 2004. The rising tide of ocean diseases: Unsolved problems and research priorities. *Frontiers in Ecology and the Environment* 2(7): 375–82.

Harvell, C.D., C.E. Mitchell, J.R. Ward, S. Altizer, A.P. Dobson, R.S. Ostfeld, M.D. Samuel. 2002. Climate warming and disease risks for terrestrial and marine biota. *Science* 296: 2158–62.

Kozloff, E. 1974. *Key to Marine Invertebrates of Puget Sound, the San Juan Archipelago and Adjacent Regions.* Seattle: University of Washington Press.

Lips, K. 2014. A tale of two lineages: Unexpected long-term persistence of the amphibian-killing fungus in Brazil. *Journal of Molecular Ecology* 23(4): 747–49.

Martel, A., A. Spitzen-van der Sluijs, M. Blooi, W. Bert, R. Ducatelle, M.C. Fisher, A. Woeltjes, et al. 2013. *Batrachochytrium salamandrivorans* sp. nov. causes lethal chytridiomycosis in amphibians. *Proceedings of the National Academy of Sciences* 110(38): 15325–29.

McCallum, H., C.D. Harvell, A. Dobson. 2003. Rates of spread of marine pathogens. *Ecology Letters* 6(12): 1062–67.

McCallum, H.I., A. Kuris, C.D. Harvell, K.D. Lafferty, G.W. Smith, J. Porter. 2004. Does terrestrial epidemiology apply to marine systems? *Trends in Ecology and Evolution* 19(11): 585–91.

McCauley, D.J., M.L. Pinsky, S.R. Palumbi, J.A. Estes, F.H. Joyce, R.R. Warner. 2015. Marine defaunation: Animal loss in the global ocean. *Science* 347, doi:10.1126/science.1255641.

Paine, R.T. 1969. A note on trophic complexity and community stability. *American Naturalist* 103: 91–93.

Palumbi, S.R., and A.R. Palumbi. 2013. *The Extreme Life of the Sea.* Princeton, NJ: Princeton University Press.

Park, D.J., G. Dudas, S. Wohl, A. Goba, S.L. Whitmer, K.G. Andersen, R.S. Sealfon, et al. 2015. Ebola virus epidemiology, transmission, and evolution during seven months in Sierra Leone. *Cell* 161(7): 1516–26.

Quammen, D. 2012. *Spillover.* New York: W.W. Norton.

Ray, C. 1988. Ecological diversity in coastal zones and oceans. In E.O. Wilson and F.M. Peter, eds., *Biodiversity.* Washington, DC: National Academies Press.

Sale, P. 2012. *Our Dying Planet: An Ecologist's View of the Crisis We Face.* Berkeley, CA: University of California Press.

Whiles, M., R. O. Hall, W. K. Dodds, P. Verburg, A. D. Huryn, C. M. Pringle, K. R. Lips, et al. 2013. Disease-driven amphibian declines alter ecosystem processes in tropical streams. *Ecosystems* 16(1): 146–57.

Zhang, S. 2017. Why are there so many more species on land when the sea is bigger? *The Atlantic,* www.theatlantic.com/science/archive/2017/07/why-are-there-so-many-more-species-on-land-than-in-the-sea/533247/.

2. CORAL OUTBREAKS

Allen, G. 2018. Battered by bleaching, Florida's coral reefs now face mysterious disease, www.npr.org/2018/05/15/611258056/battered-by-bleaching-floridas-coral-reefs-now-face-mysterious-disease.

Allen, G. R., and M. Adrim. 2003. Coral reef fishes of Indonesia. *Zoological Studies* 42(1): 1–72.

Anderson, R. M., and R. M. May. 1979. Population biology of infectious diseases. *Nature* 280: 361–67.

Andras, J. 2017. Genetic variation of the Caribbean sea fan coral, *Gorgonia ventalina,* correlates with survival of a fungal epizootic. *Marine Biology* 164: 130.

Aronson, R. B., W. F. Precht. 2001. White band disease and the changing face of Caribbean coral reefs. *Hydrobiologia* 460(1–3): 25–38.

Barker, R. 1998. *And the Waters Turned to Blood: The Ultimate Biological Threat.* New York: Touchstone.

Brazeau, D., and C. D. Harvell. 1994. Genetic structure of local populations and divergence between growth forms in a clonal invertebrate, the Caribbean octocoral *Briareum asbestinum* (Pallas). *Marine Biology* 119: 53–60.

Bruno, J., S. Ellner, I. Vu, K. Kim, C. D. Harvell. 2011. Impacts of aspergillosis on sea fan coral demography: Modeling a moving target. *Ecological Monographs* 81: 123–39.

Burge, C. A., C. J. Closek, C. S. Friedman, M. L. Groner, C. M. Jenkins, A. Shore-Maggio, J. E. Welsh. 2016. The use of filter-feeders to manage disease in a changing world. *Integrative and Comparative Biology* 56(4): 573–87.

Burge, C. A., N. Douglas, I. Conti-Jerpe, E. Weil, S. Roberts, C. Friedman, C. D. Harvell. 2012. Friend or foe: The association of Labyrinthulomycetes with the Caribbean sea fan *Gorgonia ventalina*. *Diseases of Aquatic Organisms* 101(1): 1–12.

Burkholder, J.M., A.S. Gordon, P.D. Moeller, J.M. Law, K.J. Coyne, A.J. Lewitus, J.S. Ramsdell, et al. 2005. Demonstration of toxicity to fish and to mammalian cells by *Pfiesteria* species: Comparison of assay methods and strains. *Proceedings of the National Academy of Sciences* 102(9): 3471–76.

Cohen, Y., F.J. Pollock, E. Rosenberg, D.G. Bourne. 2013. Phage therapy treatment of the coral pathogen *Vibrio coralliilyticus*. *MicrobiologyOpen* 2(1): 64–74.

Efrony, R., I. Atad, E. Rosenberg. 2009. Phage therapy of coral white plague disease: Properties of phage BA3. *Current Microbiology* 58(2): 139–45.

Ellner, S., L. Jones, L. Mydlarz, C.D. Harvell. 2007. Within-host disease ecology in the sea fan *Gorgonia ventalina:* Modeling the spatial immunodynamics of a coral-pathogen interaction. *American Naturalist* 170(6): E143–61.

Friedman, C.S., N. Wight, L.M. Crosson, G.R. VanBlaricom, K.D. Lafferty. 2014. Reduced disease in black abalone following mass mortality: Phage therapy and natural selection. *Frontiers in Microbiology* 5(78), doi:10.3389/fmicb.2014.00078.

Hallock, P., F.E. Muller-Karger, J.C. Halas, 1993. Coral reef decline. *National Geographic Research and Exploration* 9: 358–78.

Harvell, C.D. 1990. The ecology and evolution of inducible defenses. *Quarterly Review of Biology* 65: 323–40.

———. 2010. Coral reefs sending a warning signal, www.cnn.com/2010/OPINION/09/27/harvell.coral.reefs/index.html.

Harvell, C.D., W. Fenical, V. Roussis, J.L. Ruesink, C.C. Griggs, C.H. Greene. 1993. Local and geographic variation in the defensive chemistry of a West Indian *gorgonian* coral (*Briareum asbestinum*). *Marine Ecology Progress Series* 93: 165–73.

Harvell, C.D., K. Kim, J.M. Burkholder, R.R. Colwell, P.R. Epstein, D.J. Grimes, E.E. Hofmann, E.K. Lipp, et al. 1999. Emerging marine diseases: Climate links and anthropogenic factors. *Science* 285: 1505–10.

Harvell, C.D., K. Kim, C. Quirolo, J. Weir, G. Smith. 2001. Coral bleaching and disease: Contributors to 1998 mass mortality in *Briareum asbestinum* (Octocorallia, Gorgonacea). *Hydrobiologia* 460(1–3): 97–104.

Harvell, C.D., C.E. Mitchell, J.R. Ward, S. Altizer, A.P. Dobson, R.S. Ostfeld, M.D. Samuel. 2002. Climate warming and disease risks for terrestrial and marine biota. *Science* 296: 2158–62.

Harvell, C. D., J. West, C. C. Griggs. 1996. Chemical defense of embryos and larvae of a West Indian *gorgonian* coral, *Briareum asbestinum*. *Invertebrate Reproduction and Development* 30: 239–46.

Hoegh-Guldberg, O., P. Mumby, A.J. Hooten, R.S. Steneck, P. Greenfield, E. Gomez, C. D. Harvell, et al. 2007. The carbon crisis: Coral reefs under rapid climate change and ocean acidification. *Science* 318: 1737.

Hughes, T.P., J.T. Kerry, A.H. Baird, S.R. Connolly, A. Dietzel, C.M. Eakin, S.F. Heron, et al. 2018. Global warming transforms coral reef assemblages. *Nature* 556: 492–96.

Jensen, P.R., C.D. Harvell, K. Wirtz, W. Fenical. 1996. Anti-microbial activity of extracts of Caribbean gorgonian corals. *Marine Biology* 125: 411–20.

Kermack, W.O., and A.G. McKendrick. 1927. A contribution to the mathematical theory of epidemics. *Proceedings of the Royal Society of London A* 115: 700–721. Reprinted in *Bulletin of Mathematical Biology* 53(1991): 33–55.

Kim, K., and C. D. Harvell. 2004. The rise and fall of a six-year coral-fungal epizootic. *American Naturalist* 164: S52.

Kim, K., P.D. Kim, C. D. Harvell. 2000. Antifungal properties of gorgonian corals. *Marine Biology* 137: 393–401.

Lamb, J., B. Willis, E. Fiorenza, C. Couch, R. Howard, D. Rader, J. True, L. Kelly, A. Amad, J. Jompa, C. D. Harvell. 2018. Plastic waste associated with disease on coral reefs. *Science* 359: 460–62.

Maynard, J., R. van Hooidonk, C. D. Harvell, C.M. Eakin, G. Liu, B. L. Willis, G.J. Williams, et al. 2016. Improving marine disease surveillance through sea temperature monitoring, outlooks and projections. *Philosophical Transactions of the Royal Society B* 371(1689), doi:10.1098/rstb.2015.0208.

McNeil, D. G., Jr. 2017. Donors and drug makers offer $500 million to control global epidemics. *New York Times,* www.nytimes.com/2017/01/18/health /partnership-epidemic-preparedness.html.

Mydlarz, L., and C. D. Harvell. 2007. "Peroxidase activity and inducibility in the sea fan coral exposed to a fungal pathogen." *Comparative Biochemistry and Physiology, Part A* 146(1) (Jan. 2007): 54–62.

Nagelkerken, I., K. Buchan, G. Smith, K. Bonair, P. Bush, J. Garzon-Ferreira, L. Botero, P. Gayle, C. D. Harvell, et al. 1997. Widespread disease in Caribbean sea fans: II. Patterns of infection and tissue loss. *Marine Ecology Progress Series* 160: 255–63.

Onishi, Y., Y. Mohri, A. Tuji, K. Ohgi, A. Yamaguchi, I. Imai. 2014. The seagrass *Zostera marine* harbors growth-inhibiting bacteria against the toxic dinoflagellate *Alexandrium tamarense*. *Fisheries Science* 80(2): 353–62.

Raverty, S. A., L. D. Rhodes, E. Zabek, A. Eshghi, C. E. Cameron, M. B. Hanson, J. P. Schroeder. 2017. Respiratory microbiome of endangered Southern Resident Killer Whales and microbiota of surrounding sea surface microlayer in the Eastern North Pacific. *Scientific Reports* 7: 394.

Rypien, K. L., J. P. Andras, C. D. Harvell. 2008. Globally panmictic population structure in the opportunistic fungal pathogen *Aspergillus sydowii*. *Molecular Ecology* 17: 4068–78.

Sale, P. 2012. *Our Dying Planet: An Ecologist's View of the Crisis We Face*. Berkeley, CA: University of California Press.

Smith, G., L. Ives, I. Nagelkerken, K. Ritchie. 1996. Caribbean sea-fan mortalities. *Nature* 383(6600): 487.

Soffer, N., J. Zaneveld, R. V. Thurber. 2015. Phage-bacteria network analysis and its implication for the understanding of coral disease. *Environmental Microbiology* 17(4): 1203–18.

Soto-Santiago, F. J., and E. Weil. 2012. Incidence and spatial distribution of Caribbean Yellow Band Disease in La Parguera, Puerto Rico. *Journal of Marine Biology*, http://dx.doi.org/10.1155/2012/510962.

Sutherland, K., B. Berry, A. Park, D. Kemp, K. Kemp, E. Lipp, J. Porter. 2016. Shifting white pox aetiologies affecting *Acropora palmata* in the Florida Keys, 1994–2014. *Philosophical Transactions of the Royal Society B: Biological Sciences*, 371(1689): 20150205.

Tollrian, R., and C. D. Harvell. 1999. *The Ecology and Evolution of Inducible Defenses*. Princeton University Press.

Weigt, L., C. Baldwin, A. Driskell, D. Smith, A. Ormos, E. Reyier. 2012. Using DNA barcoding to assess Caribbean reef fish biodiversity: Expanding taxonomic and geographic coverage. *PLoS ONE* 7(7): p.e. 41059.

Yong, E. 2016. *I Contain Multitudes: The Microbes within Us and a Grander View of Life*. New York: Ecco.

3. ABALONE OUTBREAKS

Altstatt, J. M., R. F. Ambrose, J. M. Engle, P. L. Haaker, K. D. Lafferty, P. T. Raimondi. 1996. Recent declines of black abalone *Haliotis cracherodii* on the mainland coast of central California. *Marine Ecology Progress Series* 142: 185–92.

Ben-Horin, T., H. Lenihan, K. D. Lafferty. 2014. Variable intertidal temperature explains why disease endangers black abalone. *Ecology* 94: 161–68.

Chambers, M. D., G. R. VanBlaricom, L. Hauser, F. Utter, C. S. Friedman. 2006. Genetic structure of black abalone (*Haliotis cracherodii*) populations in the California islands and central California coast: Impacts of larval dispersal and decimation from withering syndrome. *Journal of Experimental Marine Biology and Ecology* 331: 173–85.

Corbeil, S., L. M. Williams, K. A. McColl, M. S. J. Crane. 2016. Australian abalone (*Haliotis laevigata, H. rubra* and *H. conicopora*) are susceptible to infection by multiple abalone herpesvirus genotypes. *Diseases of Aquatic Organisms* 119(2): 101–6.

Crane, M. S. J., S. Corbeil, L. M. Williams, K. A. McColl, V. Gannon. 2013. Evaluation of abalone viral ganglioneuritis resistance among wild abalone populations along the Victorian coast of Australia. *Journal of Shellfish Research* 32(1): 67–72.

Crosson, L. M., and C. S. Friedman. 2018. Withering syndrome susceptibility of northeastern Pacific abalones: A complex relationship with phylogeny and thermal experience. *Journal of Invertebrate Pathology* 151: 91–101.

Crosson, L. M., N. Wight, G. R. VanBlaricom, I. Kiryu, J. D. Moore, C. S. Friedman. 2014. Abalone withering syndrome: Distribution, impacts, current diagnostic methods and new findings. *Diseases of Aquatic Organisms* 108(3): 261–70.

Dang, V. T., K. Berkendorff, S. Corbeil, L. M. Williams, J. Hoad, M. S. Crane, P. Speck. 2013. Immunological changes in response to herpesvirus infection in abalone *Haliotis laevigata* and *Haliotis rubra* hybrids. *Fish and Shellfish Immunology* 34(2): 688–91.

Dang, V. T., K. Berkendorff, T. Green, P. Speck. 2015. Marine snails and slugs: A great place to look for antiviral drugs. *Journal of Virology* 89(16): 8114–18.

Ellard, K., S. Pyecroft, J. Handlinger, R. Andrewartha. 2009. Findings of disease investigations following the recent detection of AVG in Tasmania. *Proceedings of the Fourth National FRDC Aquatic Animal Health Scientific Conference, Cairns, Australia.*

Friedman, C. S., K. B. Andree, K. A. Beauchamp, J. D. Moore, T. T. Robbins, J. D. Shields, R. P. Hedrick. 2000. "*Candidatus* Xenohaliotis californiensis," a newly described pathogen of abalone, *Haliotis* spp., along the west coast of North America. *International Journal of Systematic and Evolutionary Microbiology* 50: 847–55.

Friedman, C. S., and L. M. Crosson. 2012. Putative phage hyperparasite in the rickettsial pathogen of abalone, "*Candidatus* Xenohaliotis californiensis." *Microbial Ecology* 64(4): 1064–72.

Friedman, C. S., M. Thomson, C. Chun, P. L. Haaker, R. P. Hedrick. 1997. Withering syndrome of the black abalone, *Haliotis cracherodii* (Leach): Water temperature, food availability, and parasites as possible causes. *Journal of Shellfish Research* 16: 403–11.

Friedman, C. S., N. Wight, L. M. Crosson, G. R. VanBlaricom, K. D. Lafferty. 2014. Reduced disease in black abalone following mass mortality: Phage therapy and natural selection. *Frontiers in Microbiology* 5(78), doi:10.3389/fmicb.2014.00078.

Fuller, A. 2017. Transmission dynamics of the withering syndrome rickettsia-like organism to abalone in California. MSc thesis, University of Washington.

Hills, J. 2007. A review of the abalone viral ganglioneuritis (AVG). Ministry of Fisheries, New Zealand.

Hollick, V. 2014. Seabed solution for cold sores. *University of Sydney News,* http://sydney.edu.au/news/84.html?newsstoryid = 13927.

Hooper, C., P. Hardy-Smith, J. Hanglinger. 2007. Ganglioneuritis causing high mortalities in farmed Australian abalone (*Haliotis laevigata* and *Haliotis rubra*). *Australian Veterinary Journal* 85: 188–93.

Lafferty, K. D., and T. Ben-Horin. 2013. Abalone farm discharges the withering syndrome pathogen into the wild. *Frontiers in Microbiology* 4(73), doi:10.3389/fmicb.2013.00373.

Lafferty, K. D., and A. M. Kuris. 1993. Mass mortality of abalone *Haliotis cracherodii* on the California Channel Islands: Tests of epidemiological hypotheses. *Marine Ecology Progress Series* 96: 239–48.

Moore, J. D., C. A. Finley, T. T. Robbins, C. S. Friedman. 2002. Withering syndrome and restoration of southern California abalone populations. *California Cooperative Oceanic Fisheries Investigations Report* 43: 112–17.

Neuman, M., B. Tissot, G. VanBlaricom. 2010. Overall status and threats assessment of black abalone (*Haliotis Cracherodii* Leach, 1814) populations in California. *Journal of Shellfish Research* 29(3): 577–86.

OIE: World Organization for Animal Health. 2018. Infection with abalone herpesvirus, chap. 2.4.1. in *Manual of Diagnostic Tests for Aquatic Animals,* www.oie.int/international-standard-setting/aquatic-manual/access-online/.

Raimondi, P., C. Wilson, R. Ambrose, J. Engle, T. Minchington. 2002. Continued declines of black abalone along the coast of California: Are mass mortalities related to El Niño events? *Marine Ecology Progress Series* 242: 143–52.

Savin, K. W., B. G. Cocks, F. Wong, T. Sawbridge, N. Cogan, D. Savage, S. Warner. 2010. A neurotropic herpesvirus infecting the gastropod, abalone, shares ancestry with oyster herpesvirus and a herpesvirus associated with the amphioxus genome. *Virology Journal* 7, doi:10.1186/1743-422X-7-308.

Steinbeck, J. R., J. M. Groff, C. S. Friedman, T. McDowell, R. P. Hedrick. 1992. Investigations into a mortality among populations of the California black abalone, *Haliotis cracherodii,* on the central coast of California, USA. In S. A. Shepherd, M. J. Tegner, S. A. Gusman del Proo, eds., *Abalone of the World: Biology, Fisheries, and Culture.* Oxford: Blackwell, pp. 203–13.

Tissot, B. N. 1990. El Niño responsible for decline of black abalone off southern California. *Hawaiian Shell News* 38(6): 3–4.

———. 1991. Geographic variation and mass mortality in the black abalone: The roles of development and ecology. PhD diss., Oregon State University.

———. 1995. Recruitment, growth, and survivorship of black abalone on Santa Cruz Island following mass mortality. *Bulletin of the Southern California Academy of Sciences* 94: 179–89.

VanBlaricom, G. R., J. L. Ruediger, C. S. Friedman, D. D. Woodard, R. P. Hedrick. 1993. Discovery of withering syndrome among black abalone *Haliotis cracherodii* (Leach) 1814, populations at San Nicolas Island, California. *Journal of Shellfish Research* 12: 185–88.

Zanjani, N. T., F. Sairi, G. Marshall, M. M. Saksena, P. Valtchev, V. G. Gomes, A. L. Cunningham, F. Dehghani. 2014. Formulation of abalone hemocyanin with high antiviral activity and stability. *European Journal of Pharmaceutical Sciences* 53: 77–85.

4. SALMON OUTBREAKS

Ackerman, P.A., S. Barnetson, D. Lofthouse, C. McClean, A. Stobbart, R.E. Withler. 2014. Back from the brink: The Cultus Lake sockeye salmon enhancement program from 2000–2014. *Canadian Manuscript Report of Fisheries and Aquatic Science* 3032: vii + 63p.

British Columbia Seafood Industry. 2014. *2013 Year in Review,* https://www2 .gov.bc.ca/assets/gov/farming-natural-resources-and-industry/agriculture- and-seafood/statistics/industry-and-sector-profiles/year-in-review /bcseafood_yearinreview_2013.pdf.

Casselman, A. 2011. Upstream battle: What is killing off the Fraser River's sockeye salmon? *Scientific American,* www.scientificamerican.com/article /what-is-killing-off-fraser-river-sockeye-salmon/.

CFIA (Canadian Food Inspection Agency). 2017. Locations infected with infectious salmon anaemia in 2017, www.inspection.gc.ca/animals /aquatic-animals/diseases/reportable/2017/infectious-salmon-anaemia- 2017-/eng/1486482527424/1486482798879.

Cohen, B. 2012. *Cohen Commission of Inquiry into the Decline of Sockeye Salmon in the Fraser River: Final Report.* Ottawa: Canada Privy Council, http://publications .gc.ca/site/eng/9.652609/publication.html.

Dannevig, B.H., and K. Falk. 1994. Atlantic salmon, *Salmo salar* L., develop infectious salmon anaemia (ISA) after inoculation with in vitro infected leucocytes. *Journal of Fish Diseases* 17(2): 183–87.

Dannevig, B.H., K. Falk, J. Krogsrud. 1993. Leucocytes from Atlantic salmon, *Salmo salar* L., experimentally infected with infectious salmon anaemia (ISA) exhibit an impaired response to mitogens. *Journal of Fish Diseases* 16(4): 351–59.

Dannevig, B.H., K. Falk, E. Namork. 1995. Isolation of the causal virus of infectious salmon anaemia (ISA) in a long-term cell line from Atlantic salmon head kidney. *Journal of General Virology* 76(6): 1353–59.

Falk, K., E. Namork, E. Rimstad, S. Mjaaland, B.H. Dannevig. 1997. Charac- terization of infectious salmon anemia virus, an orthomyxo-like virus iso- lated from Atlantic salmon (*Salmo salar* L.). *Journal of Virology* 71(12): 9016–23.

Food Safety News. 2016. Most feared salmon virus has arrived in BC waters, www.foodsafetynews.com/2016/01/most-feared-salmon-virus-has-arrived- in-bc-waters/#.WTF6XhPyu8U.

Gardner, M., and A. Bigg. 2011. *Western Canada Handbook*. Bath, UK: Footprint Handbooks.

Garver, K.A., A.A.M. Mahoney, D. Stucchi, J. Richard, C. Van Woensel, M. Froeman. 2013. Estimation of parameters influencing waterborne transmission of infectious hematopoietic necrosis virus (IHNV) in Atlantic salmon (*Salmo salar*). *PLoS ONE*, 8(12): e82296, doi:10.1371/journal. pone.0082296.

Gustafson, L.L., S.K. Ellis, M.J. Beattie, B.D. Chang, D.A. Dickey, T.L. Robinson, F.P. Marenghi, P.J. Moffett, F.H. Page. 2007. Hydrographics and the timing of infectious salmon anemia outbreaks among Atlantic salmon (*Salmo salar* L.) farms in the Quoddy region of Maine, USA and New Brunswick, Canada. *Preventative Veterinary Medicine* 78(1): 35–56.

Harris, B., S.R. Webster, N. Wolf, J.L. Gregg, P.K. Hershberger. 2018. *Ichthyophonus* in sport-caught groundfishes from southcentral Alaska. *Diseases of Aquatic Organisms* 128: 169–73.

Hershberger, P.K., J.L. Gregg, L.M. Hart, S. Moffitt, R. Brenner, K. Stick, E. Coonradt, et al. 2016. The parasite *Ichthyophonus* sp. in Pacific herring from the coastal NE Pacific. *Journal of Fish Diseases* 39(4): 395–410.

Hershberger, P.K., K. Stick, B. Bui, C. Carroll, B. Fall, C. Mork, J.A. Perry, et al. 2002. Incidence of *Ichthyophonus hoferi* in Puget Sound fishes and its increase with age of Pacific herring. *Journal of Aquatic Animal Health* 14(1): 50–56.

Hoenig, J.M., M.L.Groner, M.W. Smith, W.K. Vogelbein, D.M. Taylor, D.F. Landers, J. Swenarton, et al. 2017. Impact of disease on the survival of three commercially fished species. *Ecological Applications* 27(7): 2116–27.

Huntsberger, C.J., J.R. Hamlin, R.J. Smolowitz, R.M. Smolowitz. 2017. Prevalence and description of *Ichthyophonus* sp. in yellowtail flounder (*Limanda ferruginea*) from a seasonal survey on Georges Bank. *Fisheries Research* 194: 60–67.

Isabella, J. 2010. Meet the super sockeye. *The Tyee*, https://thetyee.ca /News/2010/06/15/SuperSockeye/.

Johnson, K. 2013. Scientists are divided over threat to Pacific Northwest salmon. *New York Times*, www.nytimes.com/2013/05/03/science/infectious-salmon-anemia-threat-divides-scientists.html.

Jones, S.R.M., G. Prosperi-Porta, S.C. Dawe, D.P. Barnes. 2003. Distribution, prevalence and severity of *Parvicapsula minibicornis* infections among anadromous salmonids in the Fraser River, British Columbia, Canada. *Diseases of Aquatic Organisms* 54(1): 49–54.

Kibenge, M.J.T., T. Iwamoto, Y. Wang, A. Morton, R. Routledge, F.S.B. Kibenge. 2016. Discovery of variant infectious salmon anaemia virus (ISAV) of European genotype in British Columbia, Canada. *Virology* 13(3), doi:10.1186/s12985-015-0459-1.

Kocan, R.M., H. Dolan, P.K. Hershberger. 2011. Diagnostic methodology is critical for accurately determining the prevalence of *Ichthyophonus* infections in wild fish populations. *Journal of Parasitology* 97(2): 344–48.

Kocan, R.M., P.K. Hershberger, T. Mehl, N. Elder, M. Bradley, D. Wildermuth, K. Stick. 1999. Pathogenicity of *Ichthyophonus hoferi* for laboratory-reared Pacific herring *Clupea pallasi* and its early appearance in wild Puget Sound herring. *Diseases of Aquatic Organisms* 35(1): 23–29.

Kocan, R.M., P.K. Hershberger, G. Sanders, J. Winton. 2009. Effects of temperature on disease progression and swimming stamina in *Ichthyophonus*-infected rainbow trout, *Oncorhynchus mykiss* (Walbaum). *Journal of Fish Diseases* 32(10): 835–43.

Kocan, R.M., S. LaPatra, J. Gregg, J. Winton, P.K. Hershberger. 2006. *Ichthyophonus*-induced cardiac damage: A mechanism for reduced swimming stamina in salmonids. *Journal of Fish Diseases* 29(9): 521–27.

Krossøy, B., I. Hordvik, F. Nilsen, A. Nylund, C. Endresen. 1999. The putative polymerase sequence of infectious salmon anaemia suggests a new genus within the orthomyxoviridae. *Journal of Virology* 73(3): 2136–42.

Lafferty, K.D., C.D. Harvell, J.M. Conrad, C.S. Friedman, M.L. Kent, A.M. Kuris, E.N. Powell, D. Rondeau, S.M. Saksida. 2015. Infectious diseases affect marine fisheries and aquaculture economics. *Annual Review of Marine Science* 7: 471–96.

LaPatra, S.E. 1998. Factors affecting pathogenicity of infectious hematopoietic necrosis virus (IHNV) for salmonid fish. *Journal of Aquatic Animal Health* 10(2): 121–31.

Lauckner, G. 1984. Diseases caused by microorganisms. Agents: Fungi. In O. Kinne, ed., *Diseases of Marine Animals*, vol. 4, part 1: 89–113. Hamburg: Biologische Anstalt Helgoland.

Lovy, J., P. Piesik, P.K. Hershberger, K.A. Garver. 2013. Experimental infection studies demonstrating Atlantic salmon as a host and reservoir of viral hemorrhagic septicemia virus type IVa with insights into pathology and host immunity. *Veterinary Microbiology* 166(1–2): 91–101.

Mapes, L. 2017. Spill of farmed Atlantic salmon near San Juan Islands much bigger than first estimates. *Seattle Times,* www.seattletimes.com/seattle-news /environment/fish-spill-bigger-than-initial-estimates-farm-destroyed/.

———. 2018. State kills Atlantic salmon farming in Washington. *Seattle Times,* www.seattletimes.com/seattle-news/politics/bill-to-phase-out-atlantic-salmon-farming-in-washington-state-nears-deadline/.

Mardones, F.O., B. Martinez-Lopez, P. Valdes-Donoso, T.E. Carpenter, A.M. Perez. 2014. The role of fish movements and the spread of infectious salmon anemia virus (ISAV) in Chile, 2007–2009. *Preventative Veterinary Medicine* 114(1): 37–46.

Margolis, L. 1991. Susceptibility of Atlantic and sockeye salmon to IHN virus in seawater. *Aquaculture Update* 55: 1–3.

Mjaaland, S., E. Rimstad, K. Falk, B.H. Dannevig. 1997. Genomic characterization of the virus causing infectious salmon anaemia in Atlantic salmon (*Salmo salar* L.): An orthomyxo-like virus in a teleost. *Journal of Virology* 71(10): 7681–86.

Morton A., and R. Routledge. 2016. Risk and precaution: Salmon farming. *Marine Policy* 74: 205–12.

Murray, A.G., R.J. Smith, R.M. Stagg. 2002. Shipping and the spread of infectious salmon anemia in Scottish aquaculture. *Emerging Infectious Diseases* 8(1): 1–5.

OIE: World Organisation for Animal Health. 2003. Infectious haematopoietic necrosis, chap. 2.1.2. in *Manual of Diagnostic Tests for Aquatic Animals,* www.oie.int/index.php?id = 2439&L = 0&htmfile = chapitre_isav.htm.

———. 2012. Independent evaluation of the OIEE reference laboratory for infectious salmon anaemia, www.oie.int/fileadmin/Home/eng/Internationa_Standard_Setting/docs/pdf/Aquatic_Commission/Evaluation_ OIE_Ref_Lab_ISA_website.pdf.

———. 2018. Infection with HPR-deleted or HPR-0 infectious salmon anaemia virus, chap. 2.3.5. in *Manual of Diagnostic Tests for Aquatic Animals,* www.oie.int/index.php?id=2439&L=0&htmfile=chapitre_isav.htm.

Okamoto, N., H. Suzuki, K. Nakase, T. Sano. 1987. Experimental oral infection of rainbow trout with spherical bodies of *Ichthyophonus hoferi* cultivated. *Nippon Suisan Gakkaishi* 53: 407–9.

Patterson, K. R. 1996. Modelling the impact of disease-induced mortality in an exploited population: The outbreak of the fungal parasite (*Ichthyophonus hoferi*) in the North Sea herring (*Clupea harengus*). *Canadian Journal of Fisheries and Aquatic Sciences* 53(12): 2870–87.

Peterson, C. H. 2001. The "Exxon Valdez" oil spill in Alaska: Acute, indirect and chronic effects on the ecosystem. *Advances in Marine Biology* 39: 1–103.

Rahimian, H. 1998. Pathology and morphology of *Ichthyophonus hoferi* in naturally infected fishes off the Swedish west coast. *Diseases of Aquatic Organisms* 34(2): 109–23.

Saksida, S. M. 2006. Infectious haematopoietic necrosis epidemic (2001 to 2003) in farmed Atlantic salmon *Salmo salar* in British Columbia. *Diseases of Aquatic Organisms* 72(3): 213–23.

Salama, N. K. G., and B. Rabe. 2013. Developing models for investigating the environmental transmission of disease-causing agents within open-cage salmon aquaculture. *Aquaculture Environment Interactions* 4: 91–115.

Scheel, I., M. Aldrin, A. Frigessi, P. A. Jansen. 2007. A stochastic model for infectious salmon anemia (ISA) in Atlantic salmon farming. *Journal of the Royal Society Interface* 4(15): 699–706.

Sindermann, C. J., and L. W. Scattergood. 1954. Ichthyosporidium disease of the sea herring (*Clupea harengus*). In C. J. Sindermann, *Diseases of Fishes of the Western North Atlantic*. Augusta, ME: Department of Sea and Shore Fisheries.

St-Hilaire, S., C. S. Ribble, C. Stephen, E. Anderson, G. Kurath, M. L. Kent. 2002. Epidemiological investigation of infectious hematopoietic necrosis virus in salt water net-pen reared Atlantic salmon in British Columbia, Canada. *Aquaculture* 212(1–4): 49–67.

Tapia E., G. Monti, M. Rozas, Á. Sandoval, Á. Gaete, H. Bohle, P. Bustos. 2013. Assessment of the in vitro survival of the Infectious Salmon Anaemia Virus (ISAV) under different water types and temperature. *Bulletin of the European Association of Fish Pathologists* 33(1): 3–12.

Traxler, G. E., and J. B. Rankin. 1989. An infectious hematopoietic necrosis epizootic in sockeye salmon *Oncorhynchus nerka* in Weaver Creek spawning channel, Fraser River system, BC, Canada. *Diseases of Aquatic Organisms* 6: 221–26.

Vidal, J. 2017. Salmon farming in crisis: "We are seeing a chemical arms race in the seas." *The Guardian*, www.theguardian.com/environment/2017 /apr/01/is-farming-salmon-bad-for-the-environment.

5. STARFISH OUTBREAKS

Bates, A. E., B. J. Hilton, C. D. Harley. 2009. Effects of temperature, season and locality on wasting disease in the keystone predatory sea star *Pisaster ochraceus. Diseases of Aquatic Organisms* 86: 245–51.

Callahan, M. 2016. Collapse of kelp forest imperils North Coast ocean ecosystem. *Press Democrat*, www.pressdemocrat.com/news/5487602–181/collapse-of-kelp-forest-imperils?artslide = 0.

Eisenlord, M. E., M. L. Groner, R. M. Yoshioka, J. Elliot, J. Maynard, S. Fradkin, M. Turner, K. Pyne, N. Rivlin, R. van Hooidonk, C. D. Harvell. 2016. Ochre star mortality during the 2014 wasting disease epizootic: Role of population size structure and temperature. *Philosophical Transactions of the Royal Society B* 371(1689), doi:10.1098/rstb.2015.0212.

Fuess, L. E., M. E. Eisenlord, C. J. Closek, A. M. Tracy, R. Mauntz, S. Gignoux-Wolfsohn, M. M. Moitsch, R. M. Yoshioka, C. A. Burge, C. D. Harvell, et al. 2015. Up in arms: Immune and nervous system response to sea star wasting disease. *PLoS ONE* 10(7), doi:10.1371/journal.pone .0133053.

Goddard, A., and A. L. Leisewitz. 2010. Canine parvovirus. *Veterinary Clinics of North America: Small Animal Practice* 40(6): 1041–53.

Gudenkauf, B. M., and I. Hewson. 2015. Metatranscriptomic analysis of *Pycnopodia helianthoides* (Asteroidea) affected by sea star wasting disease. *PLoS ONE* 10(5), doi:10.1371/journal.pone.0128150.

Harvell, C. D., D. Montecino-Latorre, J. M. Caldwell, J. M. Burt, K. Bosley, A. Keller, S. F. Heron et al. Forthcoming. Disease epidemic and a marine heat wave are associated with the continental scale collapse of a pivotal predator (*Pycnopodia helianthoides*). *Science Advances*.

Hewson, I., J. B. Button, B. M. Gudenkauf, B. Miner, A. L. Newton, J. K. Gaydos, J. Wynne, et al. 2014. Densovirus associated with sea-star wasting disease and mass mortality. *Proceedings of the National Academy of Sciences* 111(48): 17278–83.

Kohl, W. T., T. I. McClure, B. G. Miner. 2016. Decreased temperature facilitates short-term sea star wasting disease survival in the keystone intertidal sea star *Pisaster ochraceus*. *PLoS ONE* 11(4): e0153670, doi:10.1371/journal.pone.0153670.

Mah, C. 2013. Mysterious mass sunflower starfish (*Pycnopodia*) die-off in British Columbia. *EchinoBlog*, http://echinoblog.blogspot.ca/2013/09/mysterious-mass-starfish-die-off-in.html?spref = tw.

Menge, B., E. B. Cerny-Chipman, A. Johnson, J. Sullivan, S. Gravem, F. Chan. 2016. Sea star wasting disease in the keystone predator *Pisaster ochraceus* in Oregon: Insights into differential population impacts, recovery, predation rate, and temperature effects from long-term research. *PLoS ONE* 11(5): e0153994, doi:10.1371/journal.pone.0153994.

Miner, C. M., J. L. Burnaford, R. F. Ambrose, L. Antrim, H. Bohlmann, C. A. Blanchette, J. M. Engle, S. C. Fradkin. 2018. Large-scale impacts of sea star wasting disease (SSWD) on intertidal sea stars and implications for recovery. *PLoS ONE* 13: e0192870.

Montecino-Latorre, D., M. E. Eisenlord, M. Turner, R. M. Yoshioka, C. D. Harvell, C. V. Pattengill-Semmens, J. D. Nichols, J. K. Gaydos. 2016. Devastating transboundary impacts of sea star wasting disease on subtidal asteroids. *PLoS ONE* 11(10): e0163190, doi:10.1371/journal.pone.0163190.

Paine, R. T. 1969. A note on trophic complexity and community stability. *American Naturalist* 103: 91–93.

Peterson, W., M. Robert, N. Bond. 2015. The warm blob—conditions in the northeastern Pacific Ocean. *PICES Press* 23(1): 36–38.

Pfister, C. A., R. T. Paine, J. T. Wootton. 2016. The iconic keystone predator has a pathogen. *Frontiers in Ecology and the Environment* 14(5): 285–86.

Schiebelhut, L., J. Puritz, M. Dawson. 2018. Decimation by sea star wasting disease and rapid genetic change in a keystone species, *Pisaster ochraceus*. *Proceedings of the National Academy of Sciences* 115(27), www.pnas.org/cgi/doi/10.1073/pnas.1800285115.

Schultz, J. A., R. N. Cloutier, I. M. Côté. Evidence for a trophic cascade on rocky reefs following sea star mass mortality in British Columbia. *PeerJ* 4: e1980, doi:10.7717/peerj.1980.

Wares, J., and L. Schiebelhut. 2016. What doesn't kill them makes them stronger: An association between elongation factor 1-α overdominance in the sea star *Pisaster ochraceus* and "sea star wasting disease." *PeerJ* 4: e1876.

Wirtschafter, E. 2017. Scientists and fishermen scramble to save northern California's kelp forests. *KQED Science,* https://ww2.kqed.org/science/2017/01/30/scientists-and-fishermen-scramble-to-save-northern-californias-kelp-forests/.

6. NATURE'S SERVICES TO THE RESCUE

Atad, I., A. Zvuloni, Y. Loya, E. Rosenberg. 2012. Phage therapy of the white plague-like disease of *Favia favus* in the Red Sea. *Coral Reefs* 31(3): 665–70.

Barker, R. 1998. *And the Waters Turned to Blood: The Ultimate Biological Threat.* New York: Touchstone.

Briggs, H. 2018. A third of coral reefs "entangled with plastic." BBC News, www.bbc.com/news/science-environment-42821004.

Burge, C.A., C.J. Closek, C.S. Friedman, M.L. Groner, C.M. Jenkins, A. Shore-Maggio, J.E. Welsh. 2016. The use of filter-feeders to manage disease in a changing world. *Integrative and Comparative Biology* 56(4): 573–87.

Burkholder, J.M., A.S. Gordon, P.D. Moeller, J.M. Law, K.J. Coyne, A.J. Lewitus, J.S. Ramsdell, et al. 2005. Demonstration of toxicity to fish and to mammalian cells by *Pfiesteria* species: Comparison of assay methods and strains. *Proceedings of the National Academy of Sciences* 102(9): 3471–76.

Cohen, Y., F.J. Pollock, E. Rosenberg, D.G. Bourne. 2013. Phage therapy treatment of the coral pathogen *Vibrio coralliilyticus. MicrobiologyOpen* 2(1): 64–74.

Efrony, R., I. Atad, E. Rosenberg. 2009. Phage therapy of coral white plague disease: Properties of phage BA3. *Current Microbiology* 58(2): 139–45.

Fears, D. 2018. 11 billion pieces of plastic are spreading disease across the world's coral reefs. *Washington Post.*

Friedman, C.S., N. Wight, L.M. Crosson, G.R. VanBlaricom, K.D. Lafferty. 2014. Reduced disease in black abalone following mass mortality: Phage therapy and natural selection. *Frontiers in Microbiology* 5(78), doi:10.3389/fmicb.2014.00078.

Galimany, E., G.H. Wikfors, M.S. Dixon, C.R. Newell, S.L. Meseck, D. Henning, Y. Li, J.M. Rose. 2017. Cultivation of the ribbed mussel (*Geukensia demissa*) for nutrient bioextraction in an urban estuary. *Environmental Science and Technology* 51(22): 13311–18.

Greenwood, V. 2018. Billions of plastic pieces litter coral in Asia and Australia. *New York Times,* www.nytimes.com/2018/01/25/science/plastic-coral-reefs.html.

Groner, M.L., C.A. Burge, R. Cox, N.D. Rivlin, M. Turner, K.L. Van Alstyne, S. Wyllie-Echeverria, J. Bucci, P. Staudigel, C.S. Friedman. 2018. Oysters and eelgrass: Potential partners in a high pCO$_2$ ocean. *Ecology* 99, doi:10.1002/ecy.2393.

Horn, T.B., F.V. Zerwes, L.T. Kist, Ê.L. Machado. 2014. Constructed wetland and photocatalytic ozonation for university sewage treatment. *Ecological Engineering* 63: 134–41.

Jambeck, J.R., A. Andrady, R. Geyer, R. Narayan, M. Perryman, T. Siegler, C. Wilcox, K. Lavender Law. 2015. Plastic waste inputs from land into the ocean. *Science* 347: 768–71.

Keesing, F., L.K. Belden, P. Daszak, A. Dobson, C.D. Harvell, R.D. Holt, P. Hudson, et al. 2010. Impacts of biodiversity on the emergence and transmission of infectious diseases. *Nature* 468: 647–52.

Lamb, J., J.A.J.M. van de Water, D.G. Bourne, C. Altier, M.Y. Hein, E.A. Fiorenza, N. Abu, J. Jompa, C.D. Harvell. 2017. Seagrass ecosystems reduce exposure to bacterial pathogens of humans, fishes, and invertebrates. *Science* 355: 731–33.

Lamb, J., B. Willis, E. Fiorenza, C. Couch, R. Howard, D. Rader, J. True, L. Kelly, A. Amad, J. Jompa, C.D. Harvell. 2018. Plastic waste associated with disease on coral reefs. *Science* 359: 460–62.

Maynard, J., R. van Hooidonk, C.D. Harvell, C.M. Eakin, G. Liu, B.L. Willis, G.J. Williams, et al. 2016. Improving marine disease surveillance through sea temperature monitoring, outlooks and projections. *Philosophical Transactions of the Royal Society B* 371(1689), doi:10.1098/rstb.2015.0208.

McNeil, D.G., Jr. 2017. Donors and drug makers offer $500 million to control global epidemics. *New York Times,* www.nytimes.com/2017/01/18/health/partnership-epidemic-preparedness.html.

Onishi, Y., Y. Mohri, A. Tuji, K. Ohgi, A. Yamaguchi, I. Imai. 2014. The seagrass *Zostera marine* harbors growth-inhibiting bacteria against the toxic dinoflagellate *Alexandrium tamarense. Fisheries Science* 80(2): 353–62.

Raverty, S.A., L.D. Rhodes, E. Zabek, A. Eshghi, C.E. Cameron, M.B. Hanson, J.P. Schroeder. 2017. Respiratory microbiome of endangered

Southern Resident Killer Whales and microbiota of surrounding sea surface microlayer in the Eastern North Pacific. *Scientific Reports* 7: 394.

Soffer, N., J. Zaneveld, R. V. Thurber. 2015. Phage-bacteria network analysis and its implication for the understanding of coral disease. *Environmental Microbiology* 17(4): 1203–18.

Vymazal, J., and T. Březinová. 2015. The use of constructed wetlands for removal of pesticides from agricultural runoff and drainage: A review. *Environment International* 75: 11–20.

Yong, E. 2016. *I Contain Multitudes: The Microbes within Us and a Grander View of Life.* New York: Ecco.

AFTERWORD

Aronson, R. B., and W. F. Precht. 2001. Evolutionary paleoecology of Caribbean coral reefs. In W. D. Allmon and D. J. Bottjer, eds., *Evolutionary Paleoecology: The Ecological Context of Macroevolutionary Change.* New York: Columbia University Press, 171–233.

Bidle, K. D., and A. Vardi. 2011. A chemical arms race at sea mediates algal host-virus interactions. *Current Opinion in Microbiology* 14(4): 449–57.

Bolton, C. T., M. T. Hernández-Sánchez, M. Á. Fuertes, S. González-Lemos, L. Abrevaya, A. Mendez-Vicente, J. A. Flores, et al. 2016. Decrease in coccolithophore calcification and CO_2 since the middle Miocene. *Nature Communications* 10284, doi:10.1038/ncomms10284.

Coolen, M. J. L. 2011. 7000 years of *Emiliania huxleyi* viruses in the Black Sea. *Science* 333: 451–52.

Harvell, C. D., K. Kim, C. Quirolo, J. Weir, G. Smith. 2001. Coral bleaching and disease: Contributors to 1998 mass mortality in *Briareum asbestinum* (Octocorallia, Gorgonacea). *Hydrobiologia* 460(1–3): 97–104.

Johns, C., J. I. Nissimov, F. Natale, V. Knapp, H. F. Fredericks, A. Mui, B. A. S. Van Mooy, K. D. Bidle. 2018. The mutual interplay between calcification and coccolithovirus infection. *Environmental Microbiology*, doi:10.1111/1462–2920.14362.

Lamb, J., B. Willis, E. Fiorenza, C. Couch, R. Howard, D. Rader, J. True, L. Kelly, A. Amad, J. Jompa, C. D. Harvell. 2018. Plastic waste associated with disease on coral reefs. *Science* 359: 460–62.

Mojica, K. D. A., J. Huisman, S. W. Wilhelm, C. P. D. Brussaard. 2015. Latitudinal variation in virus-induced mortality of phytoplankton across the North Atlantic Ocean. *Multidisciplinary Journal of Microbial Ecology* 10: 500–513.

Pandolfi, J. M., J. B. C. Jackson, N. Baron, R. H. Bradbury, H. M. Guzman, T. P. Hughes, C. V. Kappel, et al. 2005. Are U. S. coral reefs on the slippery slope to slime? *Science* 307: 1725–26.

Tracy, A. M., M. Pilmeier, R. M. Yoshioka, S. Heron, J. Caldwell, C. D. Harvell. Forthcoming. Revisiting the elusive baseline of marine disease. *Ecology Letters*.

Ward, J. R., and K. D. Lafferty. 2004. The elusive baseline of marine disease: Are diseases in ocean ecosystems increasing? *PLoS Biology* 2(4), doi:10.1371/journal.pbio.0020120.

Index